The Sociotechnical Constitution of Resilience

Sulfikar Amir
Editor

The Sociotechnical Constitution of Resilience

A New Perspective on Governing Risk and Disaster

Editor
Sulfikar Amir
Nanyang Technological University
Singapore, Singapore

ISBN 978-981-10-8508-6 ISBN 978-981-10-8509-3 (eBook)
https://doi.org/10.1007/978-981-10-8509-3

Library of Congress Control Number: 2018935189

Cover credit: Takashi Kitajima

Printed on acid-free paper

This Palgrave Macmillan imprint is published by the registered company Springer Nature Singapore Pte Ltd. part of Springer Nature.
The registered company address is: 152 Beach Road, #21-01/04 Gateway East, Singapore 189721, Singapore

Preface

This edited volume is a peer-reviewed publication of latest works in the field of Science and Technology Studies (STS) that delve into the notion of resilience, and critically examine how it has grown rapidly in academic discourses, research projects, and policymaking. It is a collection of papers that aim to uncover the underlying social and technical factors that shape human capacity to endure crisis and dramatic changes and to recover from catastrophic events. The whole volume revolves around the concept of sociotechnical resilience, which is proposed as a novel framework to understand resilience as a hybrid construct.

The book stemmed from a workshop held at Nanyang Technological University (NTU) in June 2016. It was a two-day meeting jointly supported by NTU's Center for Liberal Arts and Social Sciences and the Future Resilient Systems, Singapore-ETH Center. I would like to thank K.K. Luke, Liu Hong, Zhou Min, and Hans R. Heinimann for their support to the workshop. Special thanks also go to Vivek Kant, Fredy Tantri, and Aaron Ang for their tremendous assistance in organizing the workshop. Introduction of this edited volume was completed during my sabbatical leave at the Holtz Center for Science and Technology Studies, University of Wisconsin-Madison for which I thank Daniel Kleinman, Samer Alatout, and Lyn C. Macgregor.

Singapore, Singapore Sulfikar Amir

Contents

Notes on Contributors

Sulfikar Amir is an associate professor of science, technology, and society in Sociology Programme, School of Social Sciences, Nanyang Technological University. He is the author of *The Technological State in Indonesia: the Co-constitution of High Technology and Authoritarian Politics*. His research interests primarily cover technological politics, development and globalization, risk and disaster, city studies, and resilience. Aside from being an academic, Amir is a documentary filmmaker. His latest film *Healing Fukushima* chronicles how a group of medical experts in Fukushima deal with radiation hazard in the aftermath of the nuclear disaster.

Charlotte Mazel-Cabasse holds a PhD of Geography and Science and Technologies Studies from the University of Paris-Est, where she studied at the Laboratoire Techniques, Territoires et Sociétés (LATTS), at Ecole Nationale des Ponts et Chaussées. She is interested in the ways in which practices and methodologies of data science transform production of knowledge and interdisciplinary collaboration, as well as scientific personae and trajectories within the academic institution. Previously a researcher at EPFL, Switzerland, she worked on research projects questioning the definition of "science," "society," "future," and "risk." She also participated to join a research-action project with UN Agencies (ISRD, WHO) in Madagascar.

Megan Finn is an assistant professor at University of Washington. Her work contributes to three key areas of research: public information infrastructures, crisis informatics, and the history of information. Her book

Documenting Aftermath, which will be published by MIT Press in 2018, is about information and communication practices after historical earthquakes in Northern California.

Stephen Healy worked for Greenpeace International in London before gaining his first academic posting in the United Kingdom. He transferred to the University of New South Wales (UNSW), via a short stint at the New South Wales Environment Protection Authority (NSW EPA), where he is currently a member of UNSW's Environmental Humanities program. His interests meander from a focus on energy and its politics through to the affective challenges posed by the Anthropocene.

Jen Henderson is a postdoctoral fellow with Western Water Assessment at the Cooperative Institute for Research in the Environmental Sciences in Boulder, Colorado. Her research focuses on issues of vulnerability across extreme weather and climate hazards. She is a founding member of the Alliance for Integrative Approaches to Extreme Environmental Events at the University of Oklahoma.

Anique Hommels is Associate Professor at the Department of Technology & Society Studies at Maastricht University in the Netherlands. She studied the role of obduracy of sociotechnology in cities (see her book *Unbuilding Cities*, 2005, MIT Press) and more recently worked on vulnerability in technological cultures (see the volume she co-edited with Wiebe Bijker and Jessica Mesman, *Vulnerability in Technological Cultures*, 2014, MIT Press).

Bingunath Ingirige holds the chair in Project Management and Resilience at the Global Disaster Resilience Centre (GDRC), University of Huddersfield, UK. He has published in the areas of resilience in the context of developing community resilience, flood adaptations, encompassing several stakeholders such as small businesses, private sector, the public sector, and NGOs. He is one of a team of experts who have compiled the 2017 Climate Change Risk Assessment (CCRA) in the United Kingdom, published every five years by the Committee on Climate Change (CCC), an independent, statutory body established under the Climate Change Act.

Kohta Juraku is a sociologist of science and technology and an associate professor at Tokyo Denki University, Japan. He finished his PhD in interdisciplinary information studies at the University of Tokyo in 2011. He has studied the social decision-making process for nuclear power utilization and radioactive waste disposal, as well as the social learning process for technological failures.

Vivek Kant is currently an assistant professor at Industrial Design Centre, Indian Institute of Technology Bombay (IITB), India. His research interests are human technology interaction (human factors and human computer interaction), systems design engineering, history and philosophy of engineering, and human knowing and acting (ecological psychology, symbolic interactionism, activity theory).

Masaharu Kitamura is President of Research Institute for Technology Management Strategy, which he founded in 2012. Previously he served as a faculty member of Tohoku University, Department of Nuclear Engineering for 36 years and now he is Emeritus Professor at Tohoku University. His professional areas include instrumentation and control of nuclear power plants, human factors and organizational safety in nuclear and general industries, and ethics in engineering. He is also active in the areas of public dialogue on nuclear risk and resilience engineering.

Anto Mohsin is Assistant Professor in Residence in the Liberal Arts Program at Northwestern University in Qatar (NU-Q). His teaching and research interests include disaster science and technology studies (Disaster-STS). His research on Indonesia's mudflow disaster has appeared in the online journal *Arcadia*.

Kurniawan Adi Saputro is a lecturer at Indonesia Institute of the Arts Yogyakarta. After completing doctoral research at Sheffield Hallam University on media audiences' response to disaster, his research continues to focus on how grassroots communications contribute to the mitigation of climate change.

Shin-etsu Sugawara is a research scientist at the Socio-economic Research Center of the Central Research Institute of Electric Power Industry (CRIEPI), Japan. He obtained his PhD in engineering at the University of Tokyo in 2012. He has worked on cross-disciplinary studies for better nuclear risk governance, with both engineering and social sciences expertise.

Makoto Takahashi is a professor at the Graduate School of Engineering, Tohoku University. He is a member of the Department of Quantum Science and Energy Engineering where he conducts research to improve safety of large-scale complex systems based on the harmonization of humans, machines, and the environment.

Justyna Tasic is a doctoral researcher at Nanyang Technological University, Singapore, and a researcher at the Future Resilient Systems, Singapore-ETH Center. She holds a master's degree in Spatial Economy from Adam Mickiewicz University, Poznan, Poland. She was formerly a specialist for urban and regional studies at the Statistical Offices in Poznan and Warsaw, Poland. Furthermore, she was a research assistant at Anton Melik Geographical Institute, Scientific Research Centre of the Slovenian Academy of Sciences and Arts, Slovenia.

Gayan Wedawatta is a lecturer attached to Aston Logistics and Systems Institute at School of Engineering and Applied Science, Aston University, UK. His research interests span across the fields of construction and disaster resilience, with special emphasis on flooding. His research areas include disaster resilience, extreme weather events and construction of SMEs, resilience of small businesses, adaptation to flooding, and sustainable construction.

List of Figures

List of Tables

CHAPTER 1

Introduction: Resilience as Sociotechnical Construct

Sulfikar Amir

What do the 9/11 terrorist attack in New York, the Fukushima nuclear meltdown in Japan, and the volcano eruption in Central Java, Indonesia have in common? Each of these catastrophic events entails massive devastation on physical infrastructures, and causes tremendous amounts of social and economic losses. Further, it spawns political ramifications that could have resulted in deep, and often dramatic, political changes. Despite differing causing agents–a Jihadist terrorist group, a failure of emergency instruments, and a natural force, respectively–and different degrees of destruction, these disasters share one trait; each is a foray against the sociotechnical system that embodies modern culture. Yet, it is this very core of our technological society that differentiates one disaster from another. For it is the sociotechnical system built and arranged in such a way to support human life that defines how a society is able to survive shocks and disturbances, be they natural or man-made. In other words, the resilience of sociotechnical systems determines whether a crisis will be short-lived or prolonged in the aftermath of disaster.

S. Amir (✉)
Nanyang Technological University, Singapore, Singapore
e-mail: sulfikar@ntu.edu.sg

S. Amir (ed.), *The Sociotechnical Constitution of Resilience*,
https://doi.org/10.1007/978-981-10-8509-3_1

Recently, resilience has become a "buzzword" of global conversation in academic and practitioner circles. Also, it is increasingly gaining currency in business corporations, policymaking, and transnational governance. Two reasons explain why resilience is emerging in prominence among top thinkers, leaders, and planners. Since entering the twenty-first century, the world has seen too many crises and extreme events that profoundly disrupted social systems, devastated infrastructures, and destabilized economies. It has never been seen in history how humanity becomes deeply vulnerable to various catastrophes with greater magnitudes and higher frequencies of occurrence. Moreover, the scale and impact of the so-called black swans (Taleb 2007) have gone beyond our capacity to cope with, despite rapid advancements in scientific knowledge produced to measure risk and predict the next disaster (Haimes 2015). Coupling with the growing trend of disaster is a slow but chronic process of environmental deformation, which prompts the global concern over climate change and long-term pollutions. This slow disaster is gradually altering the whole landscape of the Earth, affecting lands, rivers, coastlines as well as cities and urban spaces. It is an irreversible process that comes with unprecedented social, economic, and political costs (Blaikie et al. 2004; Toya and Skidmore 2007; Lindell 2013). This "cosmopolitan moment," to follow Ulrich Beck (2009), has shifted the attention of experts and leaders from managing risk of disaster to coping with its impact and consequences. It is this stream of thinking that gives rise to resilience as an enticing term in the world of disaster we live today.

In the last decade, numerous international conferences, seminars, and meetings were organized to discuss and formulate strategies and policies that are deemed practically effective and socially helpful to build disaster resilience. These conversations and dialogues address resilience as a concept that has to be well understood in order to reproduce a set of conditions that improve our abilities to endure disturbances especially at an enormous scale. Along with that, hundreds of books, edited volumes, journal articles, and scientific reports have been published to discover quintessential ingredients of resilience. In general, the chief objective of these academic works focuses on a twofold question: How to measure resilience and how to enhance it.

So why do we need another book on resilience when a plethora of publications on this subject are readily available?

Resilience is a complex construct. It arises from multi-faceted processes, and it does not stay fixed permanently. It develops and undevelops. This is the basic premise of this book. To illuminate how these resilience-making processes take place, more than one single explanation is needed. There are pre-determined factors that define how systems and communities are able to withstand unforeseen shocks. These factors are part of human intervention in the design and organization of system or community. There are also other factors that function as a series of coincidences rather than being intentionally embodied in the affected system and community. While the latter is fairly highlighted in existing works on resilience, the former is studied further because of its predictability (Holling 2001; Haimes et al. 2008). That said, the present book follows this direction while upholding a conviction that resilience is embedded in multi-layered structures; that is, the ability to recover from catastrophic damages is constituted by a multitude of system properties.

In the current literature on resilience, one can find a deep division between two major approaches; each focuses on a different domain of the underlying attributes. This division seems to follow the chasm C.P. Snow (1959) once pointed out in his famous essay on the two cultures. On one side stands a group of social scientists and humanities scholars who consider resilience as a characteristic of human societies that functions to respond to dramatic changes unfolding in the environment. This group of academics delves into how social, cultural, economic, and political circumstances shape human capability to endure bitter impacts of disaster. Thus, the notion of resilience in this perspective is inherently social, and the capacity for recovery and reconstruction lies in the arrangement of social structures and inter-personal relations. Such a view suggests that resilience resides in the way in which our society is organized, and that institutions and cultures matter a great deal. On the other side, engineers and natural scientists conceptualize resilience more as a problem of physical and material durability. Thus in this scientific approach, resilience appears to be a technical problem, which requires technical solutions. Concentrated on robustness of complex, interdependent infrastructures such as power grid, transportation, and the cybernet, these researchers are more interested in developing formulas and models to predict and improve system resilience.

While each approach has its own virtue in defining and explaining the mechanisms and processes that produce resilience, a gap is left in the middle ground between the social and technical domains that both camps scantily recognizes. It is the lack of sensibilities to what happens in between that the present book aims to elucidate. The book argues that resilience is neither merely social nor purely physical. In the modern world where technology and society are deeply intertwined in everyday life, resilience has to be understood as a double performance of social and technical systems, hence sociotechnical system. This is the focal point of this edited book. It explores resilience through a hybrid lens using the sociotechnical framework. It is posited that capacity for coping with disturbance and disruption during the time of crisis is largely shaped by how organizations, politics, and culture are entangled with technical arrangements of physical infrastructures. Describing only the social part without understanding how it is supported by the technical is deeply inadequate. Likewise, modelling the physical infrastructure without a basis for social and institutional analysis will result in weak resilience management. Thus, the chief objective the book is set to pursue is precisely to fill a lacuna in the mainstream view on resilience in which an integrated perspective of social and technical factors remains barely present. This book sheds light on how resilience is essentially sociotechnical.

Bouncing Back

Decades before it became a hype of global discussions, resilience had already been a central interest among ecologists. Back in the 1970s, resilience drew full attention of these scientists who were curious about the inner mechanisms allowing ecological systems to survive long-term disruptions caused by various kinds of natural agents (May 1972; Harrison 1979). A model was then developed by the ecologists to illustrate how the behavior of ecological system when undergoing gradual changes in response to the emergence of new environment (Holling 1973). This model revolves around a thesis that our ecological system is essentially in a constant search for a new equilibrium. In this light, ecological resilience is understood as a multi-state process of long-term transformations (Gunderson 2000). Later, researchers from this stream of inquiry began to expand the notion of resilience to include human agency in the equation. This brought the socio-ecological system into the central framework of the field, which places human activities on a par with ecological entities in the formulation of resilience (Adger 2000; Berkes et al. 2000). One of the first social scientists to bring resilience in the sociological discourse is Aaron Wildavsky (1988). In his *Searching for Safety*, Wildavsky, following Holling, highlights how resilience is remarkably important especially in a situation marked by inherent uncertainty. Despite a good deal of attention on resilience during this period, the concept remained unnoticed by academics and practitioners for years.

Then came a series of shocking events that shook up the world and marked the twenty-first century as the age of crisis. On 11 September 2001, two Boeing jet planes crashed into World Trade Center in New York City, killing nearly three thousand people and injuring more than six thousand others (Dawes et al. 2004). Not only did it change America forever, it also created a global fear of terrorism that has hitherto haunted Western civilization (Record 2003; Altheide 2006; Archick 2010). Three years later, a great earthquake with a magnitude of 9.1 erupted in the Indian Ocean. The Sumatra-Andaman earthquake instantly spawned a giant tsunami that travelled all the way to Africa affecting fourteen countries in the way. In this disaster, an estimated 283,000 people lost their lives (Lay et al. 2005). While the world was still amazed by the scale and magnitude of the Asian tsunami, Hurricane Katrina washed down New Orleans in Louisiana. The whole city was flooded for days, causing many residents to flee their homes. Not only did this unprecedented catastrophe reveal the lack of competence on the side of the Federal Emergency Management Agency (FEMA), it unleashed racial prejudices in the southern state (Stivers 2007). A few years later, another mega disaster occurred in Asia when the Tōhoku region of Japan was hit by a magnitude 9.0 earthquake, which released an enormous amount of energy resulting in a gigantic wave of tsunami. But the worst was yet to come. The tsunami struck Fukushima Daiichi nuclear power plant and, thanks to the vulnerability of the facility, it triggered a chain-reaction process leading to an unforeseen nuclear meltdown (Amir and Juraku 2014). This horrific event instantly reminded the world of the Chernobyl nuclear disaster thirty-five years before.

These disasters are just among many others; for instance, earthquakes in Haiti and Turkey, floods in London, Manila, and Jakarta, a chemical blast in Tianjin, high-speed train accidents in France, and so on. These tragic events have instigated suffering and misery to the affected people. Often times, the impacts are felt far away from the epicenter. We know that disaster is nothing new. People have dealt with many kinds of unfortunate events over time. And it is the dynamics of this particular sort of occurrences that prompted researchers from different disciplines to study disaster. Researchers were intrigued by how disaster and accident change people and how people respond to disaster and crisis. A growing body of knowledge in this area illuminate various aspects of disaster, encompassing the role of institutions and organizations (Vaughan 1999; Dovers and

Handmer 2012), community bonds (Wilson 2012), social capital and trust (Dynes 2006; Aldrich and Meyer 2015), political systems (Pelling and Dill 2010), culture (Fortun 2001; Lavigne et al. 2008), communication (Manoj and Baker 2007; Tasic and Amir 2016), physical infrastructures (Boin and McConnell 2007; O'Rourke 2007), and scientific knowledge and expertise (Knowles 2012; Button 2016).

What fascinates researchers is the fact that people recover from disaster. Regardless of the scale and magnitude, recent history shows how communities around the world possess a capacity to bounce back from horrific situations triggered by natural forces and technological failures. The process of recovery, however, varies across societies. There are crises and accidents that left prolonged impact on the people causing long-term trauma, conflicts, and other social problems unhandled. There are also those cases in which disaster affected the people in a relatively short time and the victims were able to withstand the damages and returned to normal life shortly after the shock. How and what determines the ability [or lack thereof] to recover quickly from disaster has been the central question to researchers, practitioners, and policymakers who pay most attention to resilience.

What Is Meant by Resilience?

The term "resilience" may have different meanings to different people. The general conception of resilience emphasizes the ability to recover from damage and destruction in the aftermath of a shocking event. However, what constitutes this ability varies across definitions. For instance, the U.N. International Strategy for Disaster Reduction (UNISDR) defines resilience as "the ability of a system, community or society exposed to hazards to resist, absorb, accommodate to and recover from the effects of a hazard in a timely and efficient manner." This definition is embodied in *The Hyogo Framework for Action 2005–2015,* which is an organized effort to enhance disaster resilience across the globe. A different meaning of resilience is proposed by the Intergovernmental Panel on Climate Change, which describes resilience as "the amount of change a system can undergo without changing state." In a more overarching view, Zolli and Healy (2013) propose resilience as "the capacity of a system, enterprise, or a person to maintain its core purpose and integrity in the face of dramatically changed circumstances." Manyena (2006) provides an excellent overview of resilience definition across institutions and research fields in social and natural sciences. What Manyena tells us is that different definitions of resilience reflect a very broad understanding of resilience, and indicate intellectual arbitrariness in the resilience debate. It comes as no surprise that each institution, agency, and organization defines resilience according to their interest and objectives, a transnational trend so attached to the globalization process that Kathleen Tierney (2015) has scrutinized to resemble a neoliberal project.

In the recent years, the literature on resilience has grown enormously. As noted earlier, the widespread realization of risk and crisis our society is facing of late is responsible for this explosion of academic works and publications, a growing trend that historian of disaster Scott Knowles (2016) calls "The Resilience Revolution." Every year new books and papers pop up with new insights, knowledge, and frameworks that seek to address resilience from various angles and methodological orientations. These are the results of a broad range of studies scholars, researchers, and scientists from various disciplines and institutions have achieved for better understandings of resilience. They are primarily characterized by interdisciplinary conversations (Kapucu et al. 2013) with the goal of offering a bigger picture of what constitutes resilience (Paton and Johnston 2006). In one point, they raise a fundamental question about the characteristics and patterns that exist in resilient communities and systems (Comfort et al. 2010; Zolli and Healy 2013). Another important aspect of resilience is resilience measurements. In this area, various parameters and assessment models are proposed to quantify indicators that show the extent to which communities are able to withstand massive disruption and how quickly it will bounce back (Chang and Shinozuka 2004; Cutter et al. 2008; Cimellaro et al. 2010). While measuring resilience is an important endeavor, other scholars pay more attention to unpack the processes and mechanisms that enhance resilience. From an engineering perspective, Hollnagel et al. (2007) view resilience as a paradigm shift for safety management, an orientation that is strongly influenced by complexity science. In a different perspective, social and organizational features are strongly emphasized as the core element from which the capacity for bouncing back originates (Leflar and Siegel 2013; Ross 2013). Concepts in social sciences such as social capital and trust are commonly applied to explain why some communities are more resilient than others (Aldrich 2012). Furthermore, communities, families, and grassroots groups are considered the foundation of social resilience (Ronan and Johnston 2005; Goldstein 2011). Growing concerns over climate change have also been linked to resilience, particularly urban risks that are rapidly unfolding in megacities around the world (Bosher 2008; Shaw and Sharma 2011; Alberti 2016). At the end of the spectrum of the resilience literature are those studies that seek to develop good practices and offer policy recommendations for resilience enhancement whether for communities, infrastructures, and built environments (Comfort et al. 2010; Walker and Salt 2012; Masys 2015).

Sociotechnical Resilience

A noted researcher of resilience once commented, "…resilience – the ability to recover from shocks, including natural disasters – comes from our connections to others, and not from physical infrastructure or disaster kits" (Aldrich 2017). This characterization is not fully correct. It is true that social relations particularly at the community level play a significant role in disaster resilience. However, in the modern society where social connections are deeply mediated by technical systems such as mobile phones, the Internet, and transportation, the quality of physical infrastructures significantly contributes to allowing communities to bounce back from disasters. On the contrary, there are studies that strongly emphasize the conditions of critical infrastructures as the central element of societal resilience (see for example McDaniels et al. 2008; Reed et al. 2009; Francis and Bekera 2014). These studies seem to downplay the fact that critical infrastructures are systems that highly depend on organizational and institutional capacities to withstand large-scale disturbances. While social factors are taken into account in some of these analyses, a reductionist view is prevalent on what constitutes social and institutional dynamics.

This edited volume aims to fix the shortcomings in each of the diametrical approaches above. It suggests examining resilience through a dual sphere whereby social, organizational, and institutional conditions are analyzed in conjunction with the state of technical and material durability. While the technical systems and infrastructures are integral to survivability of a society, the nature and arrangement of social structures shape the endurance and stability of physical infrastructures. It is for this reason that this book brings in the notion of sociotechnical systems, which embodies the entanglement of the social and the technical in the production of resilience.

Conceptually, the term *sociotechnical system* refers to a hybrid structure of human-machine relation. At the most micro level, this appears in the use of a smartphone by an individual user, which exemplifies a ubiquitous sociotechnical interaction that has massively transformed our behavior for the past decade. Obviously, the smartphone represents the technical, the user the social. Zooming out of the picture, we find large-scale sociotechnical systems such as transportation, telecommunication, electricity grid, and water supply network, all of which are critical infrastructures whose operation is extremely important to support our everyday life. These infrastructural systems are sociotechnical because their performance is highly dependent on individuals who are part of the organization that runs the entire system.

For decades, the notion of sociotechnical systems has been broadly observed and critically studied by an interdisciplinary field called Science and Technology Studies (STS; also known as Science, Technology, and Society). Using multiple perspectives from history, sociology, anthropology, political science, and philosophy, STS scholars have explored the culture and structure of sociotechnical systems across contexts (Hess 1995). In this light, sociotechnical systems are considered a social construction (Pinch and Bijker 1984), which means that the whole physical and technical configuration of modern technology is deeply shaped by a multitude of factors that emerge from society (MacKenzie and Wajcman 1999). Furthermore, it is argued in STS that technology does not have its own inner logic and its development does not follow a trajectory independent of social influences (Bijker and Law 1992). At the core of technological systems invariably lie networks of individuals and groups that are organized to govern these systems (Latour 1987).

The most relevant lesson to resilience the STS scholarship has provided in studying sociotechnical systems is that the construction of technology is deeply embedded in the social institutions where it exists. Yet, it does not end here. The social production of modern culture is increasingly influenced by the logics of everyday technical artefacts, such as the ATM, mobile phones, tap water, highways, and so on (Winner 1978). The bottom line is that sociotechnical systems are hybrid entities in which the social and the technical are completely fused and inseparable, producing effects that run through the entire aspects of our life (Latour 2005). They are co-produced and co-constituted (Jasanoff 2004; Amir 2012). Hence, productivity of modern society follows the mutual entanglement of both elements. Consequently, the absence or failure of either element will result in disruption.

Adopting this notion of hybridity is necessary for the present book to advance the concept of sociotechnical resilience. Rather than seeing resilience as being fully reinforced by social connections or an outcome of inherent strength of the physical element, we view resilience as an embedded feature that comes out of a hybrid construct where individuals and communities are blended with the materiality of technology. Furthermore, resilience is an emergent property produced through a process whereby the structure of social organizations and the arrangement of technical systems constantly shape one another. Accordingly, the assessment of resilience must take into account not only how each element contributes to the capacity for coping with extreme events, but also the ways in which the social and the technical are seamlessly integrated and mutually reinforced.

As a hybrid construct, sociotechnical resilience refers to the capacity of sociotechnical ensemble to transform from one configuration to another in the face of devastating force. This is what Sulfikar Amir and Vivek Kant (2018) term as transformability. Further, Amir and Kant elaborate what constitutes sociotechnical resilience in three terms, namely sociomaterial structures, informational relations, and anticipatory practices. Sociomaterial structures are the co-constitution of social and material entities in everyday activities that embody a sociotechnical system. A resilient sociotechnical system is defined by flexible sociomateriality to undergo transformation in the aftermath of disaster. Informational relations refer to the production, circulation, and utilization of information by various individuals, communities and groups, and institutions. In the time of crisis, information no doubt plays an extremely crucial role for communities and institutions to cope with uncertainties invoked by unexpected shocks. Sociotechnical resilience deals with how information is organized and managed to support coordinated emergency responses within the system and community. As informational relations constitute a key aspect of making humans and systems hybrids, it underpins organizational coordination, which requires common grounds for every system element to take concerted actions especially for disaster mitigation and recovery. Finally, anticipatory practices are a set of everyday activities and routines aimed at anticipating possibilities of uncertain situations. Anticipatory practices are informed by knowledge of potential failures and system malfunction, but also the possibilities of rapidly bouncing back from disasters to a fully operational stage. This means anticipatory practices at the levels of individuals, groups, institutions, and governments provide a head start when they are caught off guard in sudden disruptions.

Book Overview

The perspective of sociotechnical resilience has implications on structures, practices, and epistemologies that inform resilience governance. The chapters collected in this book address these aspects by illuminating a broad range of cases and stories from different parts of the world.

The book is divided into five parts. Each part touches on a different aspect, and presents real-world cases that exhibit sociotechnical resilience in different contexts. Part I explores the conceptual thinking of resilience and aims to dissect the core practices that constitute sociotechnical resilience. In her chapter, Charlotte Mazel-Cabasse describes resilience as a syncretic social process. Specifically, her chapter explores the slow emergence of sociotechnical infrastructures designed to improve earthquake risk resilience in the San Francisco Bay Area. Further, the chapter argues that resilience emerges from the collective development of a particular form of attention to the risk. In a similar vein, Stephen Healy in his chapter sheds light on how sociotechnical resilience, in its union of social and technical domains, refutes traditional knowledge precepts. Healy illustrates this tension through an examination of the roof-top photovoltaic power (PV) in the Australian electricity industry. It furnishes a case study illustrating how status quo interests may undermine resilience. Ending Part I is a chapter by Vivek Kant and Justyna Tasic that proposes a unique approach to modelling sociotechnical resilience. The model is conceptualized as a meshwork of holons that can reveal system properties. Drawing on a case study of Singapore's electricity sector, the chapter gives deep insights into system properties that defines internal conditions for sociotechnical resilience.

Moving on to Part II, two chapters demonstrate a broader scope of sociotechnical resilience in relation to the environment, and how it is tied to knowledge production and technical practices. Jen Henderson's chapter examines the interplay between scientific calculation and the ethics of accuracy that underpins the role of National Weather Service meteorologists in the United States in protecting the public from severe weather. Through the weather discourse, Henderson reveals how this sociotechnical infrastructure of risk has important social concerns due to the deployment of meteorological standards and assessment of statistical measures of predictive skill. From the other side of the globe, Anto Mohsin's chapter brings up a prolonged disaster in East Java, Indonesia. Due to an engineering malpractice, a mud volcano erupted in 2006 and it has been spewing hot mud since then, displacing thousands of people and causing massive damages to infrastructures and the environment. The chapter focuses on the affected community, which responded to the envirotechnical disaster by utilizing a platform of information sharing to fight for their safety. This platform served as a way to put up a resilient response to the seemingly unending disaster.

Part III pays specific attention to informational relations, which constitutes a key element in sociotechnical resilience. A chapter by Megan Finn brings us back to the United States to explore the way in which informational infrastructures shape resilience of American people. Finn highlights three historical examples of public information infrastructure repair following the earthquakes in Northern California. Using these examples, Finn interrogates the notion of resilience in modern disaster policy. Kurniawan Adi Saputro offers different insights from the use of social media in disaster mitigation in Yogyakarta, Indonesia when the area was struck by a volcano eruption. Saputro's chapter sheds light on collective, external responses toward mediated disaster. It follows the argument that the discursive aspect of informational relations is key to understanding collective actions in disaster mitigation.

In Part IV, our examination of sociotechnical resilience shifts to engineered systems. This part focuses specifically on the disastrous events in Japan after the 2011 Great Earthquake. Two chapters demonstrate the success and failure of sociotechnical resilience. In the chapter on a technology called SPEEDI, Shin-etsu Sugawara and Kohta Juraku apply the sociotechnical resilience framework to question why the real-time simulation technology used to predict radiation exposures in the wake of the Fukushima nuclear disaster failed to function properly, resulting in public confusion and anxiety over radiation hazards. The chapter looks into interactions and consequences of the expert notion of "precautionary action" and/or "optimization of counter measures," and questions whether these notions helped to make society more resilient. The chapter from Makoto Takahashi and Masaharu Kitamura tells an extremely important lesson from a case study when a nuclear power plant was in the brink of catastrophe. The chapter chronicles the near-disaster experience of Onagawa nuclear power station in Japan after it was struck by earthquake and tsunami in March 2011. The chapter highlights key factors that allowed the plant to avert the tragic fate of the Fukushima Daiichi nuclear power station. It draws the lessons for sociotechnical resilience that can be learned by other nuclear power operators.

Finally, Part V brings the framework of sociotechnical resilience to the urban context where spatial planning, city governance, and the economy are the core elements shaping collective capacity for coping with urban risks and uncertainties. As a whole, cities are sociotechnical systems constitutive of institutions, communities, and infrastructural networks. Hence, one can draw an overlap between city resilience and sociotechnical resilience. Bingunath Ingirige and Gayan Wedawatta elucidate city resilience in a flood-prone environment of London. Their chapter argues that businesses operation plays a significant role in terms of long-term sustainability and growth when implementing a flood resilience scheme. The last chapter by Anique Hommels brings us to the Dutch city of Rotterdam to analyze urban resilience thinking in policy documents and scholarly literature in the field of urban planning. Hommels questions the role of technology in these discourses, and probes to what extent and how resilience discourses materialize in urban planning and design choices.

References

Adger, W. N. (2000). Social and ecological resilience: are they related? *Progress in human geography*, 24(3), 347–364.

Alberti, M. (2016). *Cities that Think Like Planets: Complexity, Resilience, and Innovation in Hybrid Ecosystems.* University of Washington Press.

Aldrich, D. P. (2012). *Building resilience: Social capital in post-disaster recovery.* University of Chicago Press.

Aldrich, D. P. and Meyer, M. A. (2015). Social capital and community resilience. American Behavioral Scientist, Vol. 59 No. 2, pp. 254–269.

Aldrich, D. P. (2017). "Recovering from disasters: Social networks matter more than bottled water and batteries." *The Conversation.* February 14, 2017. Available: https://theconversation.com/recovering-from-disasters-social-networks-matter-more-than-bottled-water-and-batteries-69611. Accessed 5 October, 2017.

Altheide, D. L. (2006). Terrorism and the Politics of Fear. Cultural studies ↔ critical methodologies, 6(4), 415–439.

Amir, S. (2012). *The technological state in Indonesia: The co-constitution of high technology and authoritarian politics.* Routledge.

Amir, S. and Juraku, K. (2014). Undermining Disaster: Engineering and Epistemological Bias in the Fukushima Nuclear Crisis. *Engineering Studies*, 6(3), 210–226.

Amir, S. and Kant, V. (2018). Sociotechnical Resilience: A Preliminary Concept. *Risk Analysis*, 38: 8–16. https://doi.org/10.1111/risa.12816

Archick, K. (2010). *US-EU cooperation against terrorism.* DIANE Publishing.

Beck, U. (2009). *World at risk.* Polity.

Berkes, F., Folke, C., & Colding, J. (2000). Linking social and ecological systems: management practices and social mechanisms for building resilience. Cambridge University Press.

Bijker, W. E., & Law, J. (1992). *Shaping technology/building society: Studies in sociotechnical change.* MIT Press.

Blaikie, P., Cannon, T., Davis, I., & Wisner, B. (2004). *At risk: natural hazards, people's vulnerability and disasters.* Routledge.

Boin, A., & McConnell, A. (2007). Preparing for critical infrastructure breakdowns: the limits of crisis management and the need for resilience. *Journal of Contingencies and Crisis Management*, 15(1), 50–59.

Bosher, L. (Ed.). (2008). *Hazards and the built environment: attaining built-in resilience.* Routledge.

Button, G. (2016). *Disaster culture: knowledge and uncertainty in the wake of human and environmental catastrophe.* Routledge

Chang, S. E., & Shinozuka, M. (2004). Measuring improvements in the disaster resilience of communities. *Earthquake Spectra, 20*(3), 739–755.

Cimellaro, G. P., Reinhorn, A. M., & Bruneau, M. (2010). Framework for analytical quantification of disaster resilience. *Engineering Structures, 32*(11), 3639–3649.

Comfort, L. K., Boin, A., & Demchak, C. C. (Eds.). (2010). *Designing resilience: Preparing for extreme events.* University of Pittsburgh Press.

Cutter, S. L., Barnes, L., Berry, M., Burton, C., Evans, E., Tate, E., & Webb, J. (2008). A place-based model for understanding community resilience to natural disasters. *Global environmental change, 18*(4), 598–606.

Dawes, S. S., Birkland, T., Tayi, G. K., & Schneider, C. A. (2004). *Information, technology, and coordination: Lessons from the World Trade Center response.* Centre for Technology in Government.

Dovers, S., & Handmer, J. (2012). *The handbook of disaster and emergency policies and institutions.* Routledge.

Dynes, R. (2006). Social capital: dealing with community emergencies. *Homeland Security Affairs*, 2(2), pp. 1–26.

Fortun, K. (2001). *Advocacy after Bhopal: Environmentalism.* Disaster. New Global Orders.

Francis, R., & Bekera, B. (2014). A metric and frameworks for resilience analysis of engineered and infrastructure systems. Reliability Engineering & System Safety, 121, 90–103.

Goldstein, B. E. (2011). *Collaborative resilience: Moving through crisis to opportunity.* MIT Press.

Gunderson, L. H. (2000). Ecological resilience—in theory and application. *Annual review of ecology and systematics*, 31(1), 425–439.

Haimes, Y. Y. (2015). Risk modeling, assessment, and management. John Wiley & Sons.

Haimes, Y. Y., Crowther, K., & Horowitz, B. M. (2008). Homeland security preparedness: Balancing protection with resilience in emergent systems. *Systems Engineering*, 11(4), 287–308.

Harrison, G. W. (1979). Stability under environmental stress: resistance, resilience, persistence, and variability. *The American Naturalist*, 113(5), 659–669.

Hess, D. J. (1995). *Science and technology in a multicultural world: The cultural politics of facts and artifacts.* Columbia University Press.
Holling, C. S. (1973). Resilience and stability of ecological systems. *Annual review of ecology and systematics*, 4(1), 1–23.
Holling, C. S. (2001). Understanding the complexity of economic, ecological, and social systems. *Ecosystems*, 4(5), 390–405.
Hollnagel, E., Woods, D. D., & Leveson, N. (2007). *Resilience engineering: Concepts and precepts.* Ashgate Publishing Ltd.
Jasanoff, S. (Ed.). (2004). *States of knowledge: the co-production of science and the social order.* Routledge.
Kapucu, N., Hawkins, C. V., & Rivera, F. I. (Eds.). (2013). *Disaster resiliency: Interdisciplinary perspectives.* Routledge.
Knowles, S. G. (2012). *The disaster experts: mastering risk in modern America.* University of Pennsylvania Press.
Knowles, S. G. (2016). "We Have Learned Our Lessons: Moving Disaster Research into Practice." Keynote Address at the Workshop on Engaging Expertise in Disaster Governance. National University of Singapore, 7–8 January, 2016.
Latour, B. (1987). *Science in action: How to follow scientists and engineers through society.* Harvard university press.
Latour, B. (2005). *Reassembling the social: An introduction to actor-network-theory.* Oxford university press.
Lavigne, F., De Coster, B., Juvin, N., Flohic, F., Gaillard, J. C., Texier, P., … & Sartohadi, J. (2008). People's behaviour in the face of volcanic hazards: Perspectives from Javanese communities, Indonesia. *Journal of Volcanology and Geothermal Research*, 172(3), 273–287.
Lay, T., Kanamori, H., Ammon, C. J., Nettles, M., Ward, S. N., Aster, R. C., … & DeShon, H. R. (2005). The great Sumatra-Andaman earthquake of 26 December 2004. Science, 308(5725), 1127–1133.
Leflar, J. J., & Siegel, M. H. (2013). *Organizational Resilience: Managing the Risks of Disruptive Events-A Practitioner's Guide.* CRC Press.
Lindell, M. K. (2013). Disaster studies. Current Sociology, 61(5–6), 797–825.
MacKenzie, D., & Wajcman, J. (1999). *The social shaping of technology.* Open University Press.
Manoj, B. S., & Baker, A. H. (2007). Communication challenges in emergency response. *Communications of the ACM*, 50(3), 51–53.
Manyena, S. B. (2006). The concept of resilience revisited. *Disasters, 30*(4), 434–450.
Masys, A. (Ed.). (2015). *Disaster management: enabling resilience.* Springer.
May, R. M. (1972). Will a large complex system be stable?. *Nature*, 238, 413–414.
McDaniels, T., Chang, S., Cole, D., Mikawoz, J., & Longstaff, H. (2008). Fostering resilience to extreme events within infrastructure systems: Characterizing decision contexts for mitigation and adaptation. Global Environmental Change, 18(2), 310–318.

O'Rourke, T. D. (2007). Critical infrastructure, interdependencies, and resilience. *Bridge-Washington-National Academy Of Engineering-*, 37(1), 22.
Paton, D., & Johnston, D. M. (2006). *Disaster resilience: an integrated approach.* Charles C Thomas Publisher.
Pelling, M., & Dill, K. (2010). Disaster politics: tipping points for change in the adaptation of sociopolitical regimes. *Progress in Human Geography*, 34(1), 21–37.
Pinch, T. J., & Bijker, W. E. (1984). The social construction of facts and artefacts: Or how the sociology of science and the sociology of technology might benefit each other. *Social studies of science*, *14*(3), 399–441.
Record, J. (2003). *Bounding the global war on terrorism.* Army War Coll Strategic Studies Inst Carlisle Barracks Pa.
Reed, D. A., Kapur, K. C., & Christie, R. D. (2009). Methodology for assessing the resilience of networked infrastructure. IEEE Systems Journal, 3(2), 174–180.
Ronan, K., & Johnston, D. (2005). *Promoting community resilience in disasters: The role for schools, youth, and families.* Springer Science & Business Media.
Ross, A. D. (2013). *Local disaster resilience: Administrative and political perspectives* (Vol. 9). Routledge.
Shaw, R., & Sharma, A. (Eds.). (2011). *Climate and disaster resilience in cities.* Emerald Group Publishing.
Snow, C. P. (1959). The two cultures and the scientific revolution: The Rede Lecture, 1959. University Press.
Stivers, C. (2007). "So poor and so black": hurricane Katrina, public administration, and the issue of race. Public Administration Review, 67(s1), 48–56.
Taleb, N. N. (2007). *The Black Swan. The Impact of the Highly Improbable.* Random House.
Tasic, J., & Amir, S. (2016). Informational capital and disaster resilience: the case of Jalin Merapi. *Disaster Prevention and Management*, 25(3), 395–411
Tierney, K. (2015). Resilience and the neoliberal project: Discourses, critiques, practices—and Katrina. *American Behavioral Scientist, 59*(10), 1327–1342.
Toya, H., & Skidmore, M. (2007). Economic development and the impacts of natural disasters. Economics Letters, 94(1), 20–25.
Vaughan, D. (1999). The dark side of organizations: Mistake, misconduct, and disaster. *Annual review of sociology*, 25(1), 271–305.
Walker, B., & Salt, D. (2012). *Resilience practice: building capacity to absorb disturbance and maintain function.* Island Press.
Wildavsky, A. B. (1988). *Searching for safety* (Vol. 10). Transaction Publishers.
Wilson, G. (2012). *Community Resilience and Environmental Transitions.* Routledge, Abingdon and Oxon, ML.
Winner, L. (1978). *Autonomous technology: Technics-out-of-control as a theme in political thought.* MIT Press.
Zolli, A., & Healy, A. M. (2013). *Resilience: Why things bounce back.* Simon and Schuster.

PART I

Dissecting Resilience

CHAPTER 2

What (Sociotechnical) Resilience Is Made Of: Personal Trajectories and Earthquake Risk Mitigation in the San Francisco Bay Area

Charlotte Mazel-Cabasse

Introduction

Thinking about sociotechnical resilience invites us to return to the hybrid nature of the sociotechnical assemblages disasters make visible (Ahn et al. 2017; Amir and Kant 2018; Bennett 2004; Law 1990). Thus, this timely volume fills a gap in the field of cindynics[1] and disaster studies when it comes to understanding how emotional, physical, scientific, and technical infrastructures are intertwined with social and institutional ones in the development of resilience.

As we now know only too well, risk and disaster are hard to characterize fully, and they do resist the observer's temptation to determine a single definition or a simple explanation for either. However, how experts, scientists, residents, and the public understand risk and disaster has a considerable impact on the politics of mitigation and reconstruction. Since the end of the Second World War, a long tradition of research has

C. Mazel-Cabasse (✉)
University of California Berkeley, Berkeley, CA, USA
e-mail: charlottecabasse@berkeley.edu

S. Amir (ed.), *The Sociotechnical Constitution of Resilience*,
https://doi.org/10.1007/978-981-10-8509-3_2

explored and critically analyzed the intertwined relationships between micro- and macro-level events and the multiple stakeholders involved in the unfolding of disasters (Boudia and Jas 2007; Frickel and Bess 2007; Hoffman and Oliver-Smith 2002; Knowles 2011; Lane et al. 2010; Lash et al. 1996; Quarantelli 1998; Solnit 2009; Tierney 2001; White 1945; Wisner et al. 2006).

Despite years of research and evaluation, disasters can still surprise us (Lussault 2007), even decades after their occurrence. In the Bay Area of San Francisco, for instance, the scale of the great 1906 earthquake and fire—the "Big One"—has until recently been difficult to appreciate. A century after the fact, reopening a seemingly cold case, Stephen Tobriner, a professor of architecture at the University of California, Berkeley, showed that—contrary to the commonly accepted narrative—the 1906 earthquake did little damage when compared to the fire that followed[2] (Tobriner 2006). The discussion about the nature of the disaster (was it a fire? was it an earthquake?) and its consequences had been of capital importance at the time because of its political and economic implications (Brechin 2006; Walker 1990). But it also matters for today's science because the observation of past catastrophes has been (Coen 2013)—and still is (Juraku 2015; November and Leanza 2015)—a major source of information for scientists and experts engaged in disaster preparedness.[3]

In seismology and earthquake studies, long before the era of constant activity monitoring (Bossu et al. 2011; Taira et al. 2009), knowledge of earthquake risk came from first-hand experience and the evaluation of damages in the immediate aftermath of a catastrophe. Earthquake scientists were therefore required to show some understanding of the sociotechnical context in which the disaster happened, something that some of them continue to do, blurring the categories of expertise and lay knowledge, civic and scientific engagement (see Finn in this volume). Building on years of expertise and experience, lay experts and scientists used their intimate knowledge of the earthquake experience—how it felt, how it displaced things, how it changed the landscape—as an object for scientific inquiry. This multi-dimensional experience of the earthquake has prompted experts and scientists to take into account the experience of living through earthquakes as a valid object for science while motivating residents to better educate themselves.

As an extension of this scientific tradition, Tobriner, looking for traces of the past disaster, identified what I would argue as a key element of sociotechnical infrastructure: the collective network of individuals, citizen-observers and citizen scientists who pay attention to the risk, who look for it and prepare for it. In his account, he exposed a social network that can be described as entangled emotional and personal connections between people, things, and places: a specific form of "attachment" (Gomart and Hennion 1999) that explains why residents dedicate life-long careers to mitigating earthquake risk while simultaneously participating in the development of a thriving environment in the region they are so committed to.

During my field research in the Bay Area between 2009 and 2013, I interviewed earthquake experts (seismologists, building inspectors, urban planners, structural engineers, and others), focusing on the kind of people Tobriner referred to: people engaged locally in earthquake-risk preparedness and who are also residents of the Bay Area.[4] In Tobriner's words: "They are us: the homeowner making his home safe for his family, the investor building apartments that are safe for his family, the businessman who pays extra for earthquake safety, the inventor who deals with innumerable vexing details, the research scientist trying to understand earthquakes, the government official who implements new earthquake policies, the engineer concerned about earthquake safety who does his best to ensure it; the architect who realizes that safety is as important as architectural expression; the contractors and subcontractors who understand earthquake safety and build for it" (Tobriner 2006: 280). Many of my interviewees did, in fact, overlap in more than just one category. During these interviews, it became evident that these individuals form an epistemic community (Knorr-Cetina 1999; Mahony and Hulme 2016) entangled with the history of the development of resilient infrastructures. Their personal histories, professional trajectories, and epistemic networks have defined the contours of this sociotechnical assemblage, this network of awareness to risk—through which emerged the scientific definition of earthquake risk and the possibility of resilience.

This chapter does not discuss advances in earthquake science and engineering but rather seeks to explore the social network that connects experts and scientists to their field. To do so, I will show how the conditions for future resilience have progressively emerged, from the immediate aftermath of the 1906 earthquake until today. The chapter is built around the story of one member of this epistemic community ("Rob"), a retired architect and earthquake safety advocate whom I interviewed in 2013. Spanning several decades, his professional trajectory tells the story of the entanglement of institutions and people, actions and events that, through generations, have developed awareness and better preparedness to the earthquake risk. His recollection of successes and setbacks illustrates the plasticity, but also the fragility, of the sociotechnical infrastructure on which resilience relies. I have also quoted "Jane," whom I interviewed in 2010, a structural engineer deeply engaged in earthquake preparedness whose knowledge of the recent history of earthquake preparedness in the Bay Area helps describe the progressive constitution of resilience. Finally, I have included comments from, and references to, other Bay Area experts who participate in risk mitigation as active members of the Bay Area community.

In this piece I will argue that resilience is determined by the quality of attachment—or, to use Latour's words, "the formidable proliferation of objects, properties, beings, fears, techniques that make us do things unto others" (Latour and Girard Stark 1999). To paraphrase Latour, what is at stake here is that "what is given in experience" to scientists and earthquake experts is not "simplified too much" (Latour 2005: 2). Applying this definition to the question of disasters, I'll explore the wealth of intertwined institutions, communities, and infrastructures able to capture and cover the different spatial and temporal scales of the risk and provide for hidden interdependencies or unanticipated vulnerabilities.

Genealogy of Sociotechnical Resilience as a Process

Resilience is a slippery concept. Within a few years, it has overcome other concepts describing exposure to risk as well as individual, group, and infrastructure capacity to recover (Adger 2000; Folke 2006; Timmerman 1981). Using the concept of resilience in lieu of other existing concepts has generated some epistemic tensions, as the polysemy of the term presents both an opportunity and a heuristic challenge (Mazel-Cabasse 2017; Reghezza-Zitt et al. 2012). As a discursive and sometimes performative concept, resilience has allowed many important discussions on the hermeneutics of both disaster studies and cindynics,[5] but the multiple epistemological origins of the term also opened the possibility of contradictory and sometimes incompatible interpretations.[6] Being or not being resilient encompasses many different realities, and to ordain a form of obligation to resilience carries the possibility of stigmatization and value judgments (Bonanno 2004; Jasanoff 2014; Reghezza-Zitt and Rufat 2015; Weichselgartner and Kelman 2015). Therefore, resilience cannot be defined in advance, nor can it be requested. It can be described only after the facts, resisting the temptation of naturalization that reduces it to a simple function that would leave aside the dynamics and tensions constitutive of the social world (Comfort et al. 2010).[7]

Laying the Foundations: The First Steps Toward Resilient Infrastructures

Taking the opposing view from Mike Davis who had famously recommended letting "Malibu burn" (Davis 1998), Bay Area earthquake experts, scientists, and concerned residents have, for more than a century, worked to mitigate earthquake risk to preserve San Francisco and the surrounding cities. Rob, an architect by training, was introduced to me through a common connection who referred to him as an "elder" because of his "old professional age" as well as his feeling of being "very much a Bay Area resident and advocate" (personal communication, September 4, 2013) in the earthquake risk community. During more than three decades, Rob had been deeply involved in earthquake risk management, participating in multiple programs and initiatives, juggling positions at the local and regional levels, working with state agencies and NGOs alike, and teaching at international institutions. In his own account of this trajectory, he embraced a narrative that pays tribute to the importance of transgenerational transmission of knowledge and mitigation efforts. During our communication, he recalled his mentors:

> California was blessed with engineers and geologists, not just in the Bay Area, and created the modern science of seismology and structural engineering at Caltech, Berkeley, and Stanford. These were bigger-than-life people that included John Blume[8] at Stanford; Henry Degenkolb,[9] a structural engineer in San Francisco; Karl Steinbrugge,[10] a structural engineer who worked for the Insurance Services Organization and taught at the architectural school at Berkeley; Bob Olson,[11] a political scientist who headed the California Seismic Safety Commission for its first 20 years; Henry Lagorio,[12] an architect on the Berkeley faculty who championed designing (architectural) for seismic forces; Bruce Bolt[13] at the Berkeley Seismological Laboratory; Stanley Scott[14] at Berkeley's Institute for Governmental Studies, who pioneered seismic public policy initiatives; and the members of EERI in northern and southern California who created the seismic code provisions that eventually found their way into the Uniform Building Code. Steinbrugge, Lagorio, and Scott were my mentors and got me interested in studying earthquakes. It all happened here because there were mentors and leaders. (Rob, personal communication, October 3, 2013)

This group of very interdisciplinary scholars and engineers had a set of common interests, but it is the magnitude 9.2 earthquake that hit the state of Alaska in 1964, causing considerable physical damage and 139 casualties, that got them together. The event attracted the interest of Rob's mentors at the University of California, Berkeley. Soon, they started thinking about the kinds of damages an earthquake of this dimension could cause in the Bay Area; a monograph, *Earthquake Hazard in the San Francisco Bay Area: A Continuing Problem in Public Policy*, was published by Steinbrugge in 1968 (Steinbrugge 1968). The book drew the attention of then-California senator Alfred Alquist. The Joint Committee on Seismic Safety for the State of California, "charged with investigating existing earthquake preparedness programs and with the development of a statewide seismic safety plan" ("Inventory of the Joint Committee on Seismic Safety Records" n.d.) and chaired by Alfred Alquist, was reportedly conceived during a dinner at Berkeley's Faculty Club the following year (Chakos et al. 2002). Once these first connections between the scientists and policy makers became established, questions arose on two grounds: What are the earthquake predictions for Southern and Northern California? What should be done to avoid casualties and damage? The earthquake problem was taken seriously in the following decades, and for earthquake experts and scientists, the 1970s were a very active period, as Rob recalled:

> Little known and less remembered was the initiative of Jerry Brown, then in his first two terms as governor in the late 1970s and early 1980s, in the creation of the Governor's Earthquake Taskforce—comprised of 400 "experts" on 40 committees that were tasked to assess the risk and response capacity of the state and local governments and to develop plans to fill the gaps. The governor directed the state geologist, James Davis, to create scenarios for the most likely earthquakes in southern and northern California [in order to] enable the task force committees to focus on the impacts of the earthquakes. These (eventually seven-plus) scenarios were pioneering efforts to quantify the impacts of the ground motions on structures, lifelines, infrastructure, hospitals, schools, etc. and painted a picture story of what an earthquake like [that of] 1857 or 1906 would be like in today's physical environment. (Rob, personal communication, September 3, 2013)

Earthquake experts have long been advocates of regulatory solutions to minimize risks caused by unsafe buildings. Structural engineers, urban planners, and seismologists are very active in the adoption of state laws that require California to construct buildings that are more earthquake-resistant, specifically through zoning, housing policy, good governance, and infrastructure policy (ABAG 2013).

The Materialization of Resilience

The process by which seismic provision, guidelines, regulations, and building codes have been defined and continue to be constantly revisited, is a good illustration of the progressive materialization of the network of awareness into pragmatic measures. It also shows the long-term emergence of resilience in large urban areas.

Here again, earthquakes prompted the introduction of regulatory frameworks; the first one was the magnitude 6.4 Long Beach earthquake of March 10, 1933, that badly hit Southern California. Although damages were limited, they surpassed the admitted level of destruction for an event of this magnitude, especially in schools. Newspapers relayed the outrage of parents. Harry Wood, then director of the Seismological Laboratory at the California Institute for Technology, made his point very clear to newspaper reporters hoping to get some traction:

> The practical lesson of the recent temblor, as of all others, [...] is built well and choose or prepare strong foundations. Design for strength and [construct] conscientiously, using good materials. Avoid what experience has shown to be faulty. (Wood quoted in Geschwind 2001: 107)

The Long Beach earthquake can be considered a starting point of a modus operandi that has since then defined the collaboration between organizations dedicated to risk and earthquake preparedness: joint committees working collaboratively and moving progressively from the local level to the state level, then up to the federal government. In the following years, discussions were held which ultimately led to both the Field Act[15] and the Riley Act,[16] two laws that focused on the safety of public buildings and attempted to avoid proliferation of unreinforced masonry buildings.[17] Addressing what had become a common concern in California, the first seismic code, the Building Code of California, was published in 1939[18] after more than 10 years of work by small committees.[19] The expanding need for building regulations in the Bay Area in the 1940s[20] resulted from a real estate boom throughout the state.

Building codes were later reinforced in 1970 and 1990. Until the 1970s, it was the norm that local authorities published their own building codes. The California Building Standards Code (California Code of Regulations, Title 24) was created only in 1978 as an amalgamation and reorganization of existing codes. After the San Fernando Valley earthquake in 1971, Alquist introduced in 1972 the Alquist-Priolo Earthquake Fault Zoning Act prohibiting "the siting of most structures for human occupancy across traces of active faults that constitute a potential hazard to structures from surface faulting or fault creep"[21] (Alquist-Priolo Earthquake Fault Zoning [AP] Act 2012). Also in 1972, then California governor Ronald Reagan established the Governor's Earthquake Council, which was followed in 1974 by the Seismic Safety Act—the bill that created the Seismic Safety Commission.[22] Jane, a structural engineer, recalled how the concerns over building safety progressively moved from safer schools to safer hospitals to the idea that some buildings welcoming the public should be reinforced so as not to be life-threatening, but rather what she called "life-safe."

> [After] the 1971 San Fernando earthquake, the State of California realized that hospitals are facilities that really should be standing up after an earthquake. And they transferred the design of hospitals to an agency called OSHPD—Office of Statewide Health Planning and Development.[23] Every hospital has to go through a plan check and a design check by OSHPD. New hospitals are doing pretty well, but we still have this huge inventory of hospitals designed before. Finally, 1989 happens, the Loma Prieta earthquake, and once again, people are thinking, "Okay, our hospitals are still vulnerable … ," and so they started to—this is just the State of California—create a group that looked into a Senate bill. They eventually passed SB1953[24] and came up with deadlines for when you need to evaluate your hospital, when you need to have it "life-safe," and when you need to have it meet full code. And this was 20-odd years after the 1971 earthquake. So, if you look at [it] from start to finish, that's a 50-year process, almost. It's huge. But that came from the top down. And the reason that it takes so long is that it's outrageously expensive, it's billions of dollars. (Jane, personal communication, 2010)

In 1986, the State of California passed the "Unreinforced Masonry Building Law" (SB 547), which required each jurisdiction in California's "seismic hazard zones" to reduce unreinforced masonry hazards.[25] For earthquake preparedness within a community, materializing resilience into generalizable regulation remains a complicated process involving the design of seismic provisions into law and establishing guidelines that will

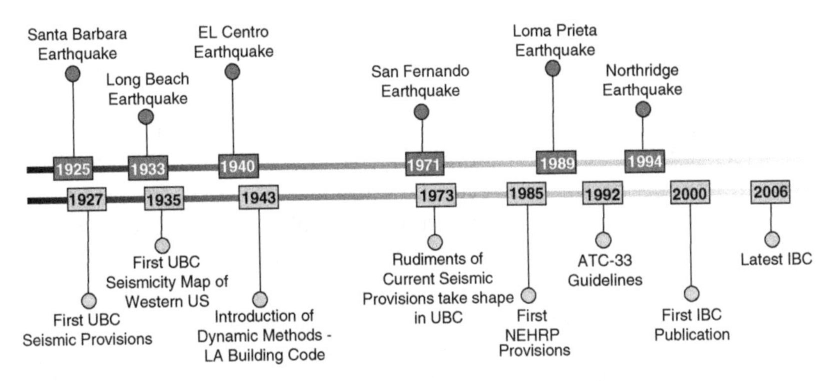

Fig. 2.1 Progression of seismic design provisions in building codes following significant California earthquakes since 1925. (Based on Financial Management of Earthquake Risk, Earthquake Engineering Research Institute, 2000 in Grossi and Muir-Wood 2006: 11) (The present scheme focuses on the evolution of the building codes)

later be implemented in the field. Since 1933, more than 80 laws have been passed in California to promote earthquake safety and mitigate damage and injury. As Fig. 2.1 shows, major progress has often been made after devastating events.

Never Again: The Trauma That Comes First

The time between two earthquakes is often a time of relief, when earthquake risk falls under the radar and experts and legislators don't get much attention. In California, things changed again in 1989, when the Loma Prieta earthquake hit the Bay Area and killed 63 people. Suddenly, the possible had taken concrete form; the simulation scenarios that had remained abstract speculations until that point were getting real.

The Cypress Street Viaduct, part of the I-880 (also called the Nimitz Freeway), used to cross through low-income neighborhoods in Oakland. On October 17 of that year, it partly collapsed, the upper tier falling on top of the lower one and resulting in 42 fatalities. The event and its aftermath were broadcast widely, showing the extremely difficult working conditions of the rescue teams and graphic images of the consequences of the catastrophe on the structure of the bridge—including crushed cars and human bodies. These images were shared by many: neighbors, families, friends, and (of course) a huge and distant television audience.[26] The large community of witnesses had to deal with direct or indirect memories of this terrible event. The emotional consequences for the rescue teams and the victims have been largely documented (Bourque and Russell 1994; Nigg and Mileti 1998; Tubbesing and Mileti 1994), as some of those involved have endured distressing and recurrent memories of the experience.

During the rescue work on the collapsed highway, firefighters worked in the pancaking space between the upper and lower levels of the freeways. [...] There were strong aftershocks and subsequent compression of the levels. As the space compressed, the escape routes became smaller, increasing the risk that workers would become trapped. Gasoline was leaking from crushed vehicles, increasing the risk of explosions if an extrication tool caused sparks. Personnel were exposed to traumatic stimuli, hearing cries for help become dim as time passed, or witnessing extraordinary procedures such as necessary limb amputation with chainsaws to extricate one of the persons trapped (Myers and Wee 2005: 185).

For the local population, the shock was terrible; numerous accounts of the events were circulated, sharing the pain, the horror, and the shock (Fig. 2.2).

Fig. 2.2 Cypress Street Viaduct in Oakland, from the 1989 Loma Prieta earthquake. (USGS photo. Not dated. Probably late 1989. H.G. Wilshire, U.S. Geological Survey)

> That Cypress Freeway, as that overpass was called, was a monumental structure in the most literal sense, a gray mass of concrete as high as an office building. When I first got there in the early evening, I remember also that I looked up and felt dizzy and sick, as though I were staring at a body that had been disemboweled. The ripped freeway had opened a great cross-section of gigantic construction innards that sprawled and jutted and smoked and hissed. Parts of cars were visible inside the striations of concrete and metal, but the whole arrangement made no sense; you couldn't understand what was roadway and what was broken piling, and on the ground were cars upside down and jerked over at weird angles between the fire trucks and improvised emergency equipment, and at one spot below the overpass a crowd of medical workers had gathered at the base of a long ladder that stretched right up into the mess overhead. (Gorney 2009)

Acknowledging their own trauma as well as their deep sense of grief and responsibility, Bay Area experts were determined not to lose the momentum—the ill-named "window of opportunity" that comes after a large disaster. Community organizing became important, and a lot needed to be done at every level, from the individual up to the state.

Emotional Bonding and Politics: The Co-construction of Resilience

Concerned residents—homeowners, businessmen, inventors, research scientists, government officials, engineers, architects, contractors—that Tobriner had mentioned came together, as did community organizers, local officials, building departments, emergency managing offices, churches, and schools, to determine what a safer Bay Area could look like. Inspiration about how to get things done did come from the field community organization: In the 1960s Rob had studied at UC Berkeley and been inspired by the work of Saul Alinsky, a Chicago native and community organizer who had written the book *Rules for Radicals* (Alinsky 1989).[27] At a moment when the modality of resilience and the infrastructure that will support it needed to be invented, Alinsky's work made a difference, as Rob explains:

> We wanted to empower people. We wanted them to feel they [had] the control of the situation. We would go from city to city and talk to the building department people, show them our scripts and propose to talk to their supervisor or to city council. We offered them a "planning partnership"

> which helped them identify the buildings, infrastructures, and systems at risk and what could be done to improve their organization and their capacity to face a disaster. (Rob, personal communication, October 3, 2013)

Focusing on a very different approach than the ones favored at the time, his team brought together scientists and senior political staff to explain what an earthquake would look like for each institution that they worked with, including "a Southern California county (San Bernardino County), a large city (Los Angeles), a small city (Westminster in Orange County), and a corporation (Security Pacific Bank)" (Geschwind 2001: 217).

As Rob recalled, these meetings included a lot of coffee and a lot of listening, even more so than talking. The team motto was "do not focus on *you should*." Instead, these meetings provided a safe environment to think about the "unthinkable"—the injuries, deaths, and financial devastation that a large catastrophe could provoke. In each of these meetings, the group members act as a translator for government personnel and for private institutions, "making sense of the language of engineers" (Chakos et al. 2002: 7) with a very strategic perspective:

> For each of these institutions, project staff members worked together with emergency preparedness personnel to develop detailed response plans and contingency plans for a large earthquake. Once these prototype plans had been worked out, they could then be transferred to similar entities across the region. In this manner, earthquake planning increased significantly in Southern California, because individual preparedness offices now had detailed models of emergency plans [...] available to them. (Geschwind 2001: 217)

Within a few years, the team was involved in shaping earthquake preparedness plans for San Francisco, Berkeley, Oakland, Santa Cruz, Palo Alto, and Santa Barbara. In each of these cities, they focused on what they termed "sweet spots": schools, hospitals, and government buildings. In just a decade, Berkeley residents voted for several measures that collectively allowed more than $390 million from local taxes to be used for municipal safety improvements including seismic retrofitting of city buildings and various school safety programs[28] (Chakos et al. 2002). Some have said this was one of the most important "self-inflicted" increases in local taxes in the history of the United States.

Finally, following the 1989 earthquake, large public and infrastructure works began, which made the potential risks even more visible for residents of the Bay Area. As one respondent commented:

> The concern that the Bay Area's residents have had since the Loma Prieta earthquake with regard to infrastructure is visible. They've seen the signs, gone through the disruption, seen the toll bridges and interchanges on highway bridges retrofitted. Their bills for water [were] increased to pay for upgrades to the water system. They've seen the fares and tolls and their property checks and bills change to pay for the retrofitting of the BART system. They've seen [that] this not only hit them personally, in terms of money, but they've also seen the construction happening. I think there is a larger group of people who understand they have to retrofit their homes than used to be the case. (Jane, personal communication, 2010)

There is no doubt that these actions were made possible by the social and political context in which they occurred. For several decades, Berkeley, and the larger Bay Area, had been known for its strong attachment to a lively democratic tradition, even within the walls of city hall. In this "atmosphere in which leaders take responsibility for reducing risk" (Chakos et al. 2002), the work of this small team reached out to more communities, and many cities started to take concrete, visible actions. Does this mean that the Bay Area is now ready for the next Big One? Things are, of course, more complicated.

Fragile Infrastructure, Fragile Resilience

One ricochet effect of the 9/11 catastrophe was the nationwide reorganization of 22 separate disaster-mitigation agencies under the supervision of the Department of Homeland Security (DHS). At the time, these agencies accounted for 180,000 employees and 650 separate computer systems, according to Charles Perrow (Perrow 2006). "The new department merged agencies that, along with their security roles, had responsibilities for such activities unrelated to terrorism as fisheries, river floods, animal diseases, energy reliability, computer crime, citizenship training, tariffs on imports, drug smuggling, and the reliability of telephone networks" (Perrow 2006). For many disaster-management experts, this "potpourri of unrelated activities" was not helpful in reducing terrorist threats and "may even have reduced our protection from our two other types of disaster, natural and industrial" (Perrow 2006).[29]

These political changes impacted the local network of earthquake experts: "It was an overnight change," recalled Rob. Within months, his office library in Oakland was closed down and 20 years' worth of documentation and archives was thrown away. In 2003, Arnold Schwarzenegger took office as governor of California; Rob recalls: "He said to the people of California, 'If you need something, call me!'—which doubled the amount of work." Of course, these calls largely disturbed the organization of the group, forcing them to answer many kinds of solicitations, even those not related to earthquake preparedness. The expertise and dedication of the group were no longer considered paramount, and its effectiveness became diluted by the arcane world of administration. As time passed with no reminders of the region's seismic activity, the earthquake problem became a low priority on the political agenda: for more than a decade, the message for local governments was essentially "*You're on your own.*" Experts worried about a total disengagement of state and federal government support, hoping that the actions taken during the 1990s would suffice to limit the effects of major catastrophes, including any large-scale earthquakes (Chakos 2006).

In addition, and despite all these efforts, research was showing that the building codes, as good as they could be, were often not enforced by local building departments (Burby and May 1998). After more than a century of building resilient infrastructure, "*The greatest challenges to California's building departments continue to be: 1) The lack of public awareness and political support for effective code enforcement; 2) The risks of existing vulnerable buildings in earthquakes; and 3) The need to enhance staff sizes and qualifications to keep up with phenomenal growth.*" (Turner 2004: 10) reminded structural engineer Fred Turner, another important figure in the Bay Area's earthquake community.

Conclusion

Today, after 30 years of working in the field of disaster preparedness in the Bay Area, Rob has retired, and the task of providing information and advocacy is pursued by a new generation of experts working hard to maintain the level of awareness necessary to good earthquake preparedness. Since Hurricane Katrina, Bay Area earthquake experts have learned their "ABK" ("All But Katrina"), and the turn of the century has brought renewed interest in the question of earthquake safety. Building on their existing network and surfing the wave of new and potentially game-changing projects such as Earthquake Early Warning systems (Lin 2017; Strauss and Allen 2016), they still rely on that strong community to integrate new technological systems and reinforce the strengths of the interactions and dependencies of each urban system. They acknowledge that multiple sociotechnical systems will come into play: local governments, schools, water, transportation, churches, private companies, gas, communications, electricity, health and human services, and so on.

Experts' continuous efforts to understand the mechanisms of earthquakes and their consequences for built environments have created and developed further the regulatory apparatuses designed to strengthen building construction codes in California, and to avoid—as much as possible—major damage caused by an earthquake.

> The story of earthquake hazard mitigation in twentieth-century California, then, is the story of the progressive impulse among a small group of California scientists and engineers and the persistence of that impulse even after other reformers had become disenchanted with the state. (Geschwind 2001)

Through the recollections and personal trajectory of experts, this chapter has proposed a situated view on the social and historical conditions that led to the emergence of sociotechnical resilience in the San Francisco Bay Area. It is the experience of living with the earthquake—of waiting for it, fearing it, remembering it, and getting ready for the next one—that gives sense to these complex sets of actions, provides a space for the sociotechnical resilience to emerge, and defines the contours of a collective space of risk: a world composed and organized around constantly moving tectonic plates. The exploration of Rob's history also tells us another story: that of the fragility of the network of risk awareness, which can at any time be weakened by the wrong kind of attachments (Latour and Girard Stark 1999).

In this chapter, I have addressed resilience as a syncretic social process and have focused on one of the often-overlooked dimensions of the constitution of resilient infrastructure: the strength of the network of awareness that supports the sociotechnical infrastructure. Such an examination shows the slow and progressive co-construction of scientific and lay knowledge, building codes, public information, progress in structural engineering, and, later, the more systematic cross-evaluation of independency and path dependency that might have created the conditions for resilience in the Bay Area. By focusing on the sociotechnical infrastructures that make resilience possible, this chapter is an invitation to think about resilience not as a generic set of tools and good practices, but as a situated endeavor whose success and failure cannot be attributed to an individual actant alone but to a chain, a network with its own historiography, its own geographical and epistemological dimensions.

In light of recent work by Peter Hall and Michèle Lamont, I want to point to the importance of the politics of this network and its consequence on the different modalities of resilience. Exploring the concept of social resilience, Hall and Lamont noted that it "depends not only on the feature of society on which individuals draw to enhance their capabilities but also on the capacity of communities to mount collective response to challenges." Moving from the individual scale to the more collective one, social resilience can be defined by "the capacity of groups of people bound together in organizations, classes, racial groups, communities, or nations to sustain and advance their well-being in the face of challenges" (Hall and Lamont 2013: 2). Taking into account the transformability dimension (Amir and Kant 2018) of the disaster, there remains the question of the large variability of vulnerability—the cultural, economic, and social repertoires (Hall and Lamont 2013: 14) that different groups are able to mobilize—within the same space.

The considerable work in education and outreach conducted by experts such as Rob has relied on the idea that tailored recommendations gleaned from constant field work and from an in-depth knowledge of each community's vulnerability will create better preparedness and, in the long term, better resilience. But more granular research needs to be done to reconnect this in the fields of disaster studies and cindynics.

Notes

1. The science of risk analysis.
2. Looking at pictures taken directly after the earthquake but before the fire, Tobriner established a new estimation of the 1906 destruction. In doing so, he also contested what *"many scholars and popular historians have accepted and repeated, [namely] the idea that San Francisco of the 1860s denied the existence of seismic danger [...]. However, historical records show that architects, engineers, and even everyday citizens understood the consequences of the earthquake of the 1860s and tried to inventory the damages, to*

understand what had happened, retrofit buildings to resist future earthquakes and to build earthquake resistant structures" (Tobriner 2006: 35).

3. In 2006, scientists and experts modeled and analyzed the possible impacts of an earthquake similar to the 1906 Big One in today's San Francisco (Perkins et al. 2006).
4. Excluded from the interview process was the insurance and actuarial sector, which is covered by other, ongoing research (Johnson 2011).
5. The science of risk analysis.
6. "Resilience has been used by the physical sciences to describe the capacity of an object to bend under pressure and go back to its original form. It also has been largely utilized by ecologists to define the capacity of a system to return to a state of equilibrium after a destructive event. These interpretations might prove useful for disaster management, emphasizing the positive side of vulnerability and providing a desirable future to lean toward" (Guarnieri 2017).
7. "Some cities do better in the face of disaster than others. It is tempting to describe apparent success in terms of resilience and apparent failure in terms of a shopping list of explanatory variables. Resilience then becomes the synonym for survival and the prescribed antidote for administrative shortcomings. This is too simple. … Far from a fix-it-and-forget-it approach, resilience is the outcome of a long-term process; enduring resilience is a balancing act between risk and resources, between vulnerabilities and escalating or unmanageable risk" (Comfort et al. 2010: 272–273).
8. John A. Blume (1909–2002) was considered by many of his peers as the "father of earthquake engineering" (Blume n.d.).
9. Henry J. Degenkolb (1913–1989) served on the President's Task Force on Earthquake Hazard Reduction and received the American Society of Civil Engineers Ernest E. Howard Award (Contributors n.d.-b).
10. Karl V. Steinbrugge (1919–2001) was a professor of structural design at the University of California, Berkeley. One of his students recalled he had been very influential in defining earthquake risk: "For many of us in the design and construction fields, Karl Steinbrugge was the first person to explain the impact that earthquakes would have on our professional lives. He was quite passionate about the subject in clearly affirming that earthquakes needed to be better understood. In order to make improvements in engineering, planning, public policy, and emergency response, you had to learn from the past. As I think about the issues that we face today, I realize that what Karl was so passionate about has stayed with me long after the classes were over" (Steinbrugge n.d.).
11. Robert A. Olson, president of Robert Olson Associates, Inc., a consulting firm focusing on emergency management, hazard mitigation, community recovery planning, research, and knowledge transfer (Olson n.d.).

12. Henry J. Lagorio (1923–2013), a professor of architecture and former dean of the College of Environmental Design, was known for "chasing earthquakes," his study of architectural design for earthquakes, and his loss-estimation methods (Lagorio n.d.).
13. Bruce Bolt (1930–2005) was a professor of earth and planetary sciences at the University of California, Berkeley and a pioneer of engineering seismology (Contributors n.d.-a).
14. Stanley Scott was John A. Blume's student and later became a seismic-safety policy analyst with the Institute of Governmental Studies at the University of California, Berkeley. He recalled, "Blume was convinced that in order for buildings to withstand severe earthquake loading, both elastic and inelastic ranges of motion had to be understood and considered in design. This was a revolutionary theory that Blume would continue to refine and push for inclusion in building codes and engineering design practice for the next fifty years" ("Academic contribution: John A. Blume Earthquake Engineering Center" n.d.).
15. Assemblyman C. Don Field, a Republican, prepared the Field Act. For Tom Turner, "The Field Act transferred the regulation of public school design and construction from local governments to the State's Division of Architecture. Its enforcement on over 70,000 school construction projects since then has generated significant improvements in the practice of earthquake engineering" (Turner 2004: 4).
16. Approximately 10–15 percent of California's buildings were constructed before 1933, at a time when few cities had building codes. The Riley Act "requires local governments to have building departments that issue permits for new construction and alterations to existing structures and conduct inspections. Permit fees paid by building owners generally fund the work of local building departments. The Act also set minimum seismic safety requirements that have since been incorporated into all building codes." Western States Seismic Policy Council. (www.wsspc.org/policy/California.shtml).
17. *"An unreinforced masonry building (or UMB, URM building) is a type of building where load-bearing walls, non-load-bearing walls or other structures, such as chimneys, are made of brick, cinderblock, tiles, adobe or other masonry material that is not braced by reinforcing beams. The term is used in earthquake engineering as a classification of certain structures for earthquake safety purposes and is subject to minor variation from place to place. URM structures are vulnerable to collapse in an earthquake. One problem is that most mortar used to hold bricks together is not strong enough. Additionally, masonry elements may "peel" from the building and fall onto occupants or passersby outside"* (Contributors n.d.-c).

18. A first edition of the Uniform Building Code was published in 1927. The eastern part of the United States has had building codes in place since 1914.
19. It is interesting to note that committees remain the principal form of organization for structural engineers and the seismic community in general. Most of the reports produced in recent years have been done via committee consensus.
20. The Vesano Building Code, named after Harry Vesano, former director of Public Works in San Francisco, was published in 1948.
21. He also worked on the Alfred E. Alquist Hospital Facilities Seismic Safety Act of 1983.
22. Now the California Seismic Safety Commission.
23. Office of Statewide Health Planning and Development.
24. SB 1953 was, in fact, written in 1994 after the Northridge earthquake as the continuation of the Alfred E. Alquist. Also known as the "Hospital Act." *"The Alfred E. Alquist Hospital Seismic Safety Act ("Hospital Act") was enacted in 1973 in response to the moderate Magnitude 6.6 Sylmar Earthquake in 1971 when four major hospital campuses were severely damaged and evacuated. Two hospital buildings collapsed killing forty-seven people. Three others were killed in another hospital that nearly collapsed. In approving the Act, the Legislature noted that: '[H]ospitals, that house patients who have less than the capacity of normally healthy persons to protect themselves, and that must be reasonably capable of providing services to the public after a disaster, shall be designed and constructed to resist, insofar as practical, the forces generated by earthquakes, gravity and winds.' (Health and Safety Code Section 129680) When the Hospital Act was passed in 1973, the State anticipated that, based on the regular and timely replacement of aging hospital facilities, the majority of hospital buildings would be in compliance with the Act's standards within 25 years. However, hospital buildings were not, and are not, being replaced at that anticipated rate. In fact, the great majority of the State's urgent- care facilities are now more than 40 years old"* (California Seismic Safety Commission 2001: 3).
25. These structures are, still today, among the most hazardous buildings in the event of an earthquake.
26. *"I was living in Maine. I once lived in San Francisco and was galvanized by the images on TV of familiar places in ruins. So yes, shock waves were far-reaching. San Francisco is one of those "national cities" that everyone identifies [with]—had it been someplace like Topeka or Omaha, the images of ruin would not have carried the shock of recognition in the rest of the country,"* recalls a respondent (John, interviewed in 2009).
27. Originally active in the labor movement, Alinsky had worked in some of Chicago's poorest neighborhoods in the late 1930s and rapidly become an

inspiration for activists and politicians alike, such as Barack Obama. Inspired but also criticized, Alinsky's followers redefined the objectives and methodologies of community organizations and mobilization.

28. In four local elections, Berkeley voters approved more than $390 million in local taxes to fund mitigation projects. In addition, the city now rebates one-third of its real estate transfer tax, up to a maximum of $1500, for seismic retrofit—Goldfarb was the swing vote in agreeing in 1991 to increase the transfer tax only if the city council agreed to the rebate. As a result, more than 39 percent of Berkeley's 22,000 single-family residences and more than 30 percent of small multi-family buildings now have improved seismic resistance, at a cost to the city of more than $10 million in foregone taxes. The Disaster Council—like the Seismic Safety Commission at the state level—has kept earthquake safety on the agenda of the city council (Chakos et al. 2002).
29. A situation conducive to the disastrous response to Hurricane Katrina in 2005 (Comfort 2006; Frickel and Bess 2007). Perrow noted: "While DHS promulgates an 'all-hazards' approach, Hurricane Katrina in 2005 prompted inquiries that disclosed substantial funds were diverted from programs aimed at natural disasters to those focused on potential terrorist attacks. First responder funds, for example, were cut. Funds for anti-terrorist efforts (improved documentation requirements; watch lists; surveillance of mosques, ports, airports and public buildings; and many of the disturbing provisions of the PATRIOT Act) do not always help with the other hazards" (Perrow 2006).

References

ABAG. (2013). *Resilience Initiative- Building a Disaster Resilient Bay Area.* Retrieved from http://quake.abag.ca.gov/projects/resilience_initiative/

John A. Blume Earthquake Engineering Center. (n.d.). Retrieved January 1, 2017, from https://quake06.stanford.edu/centennial/tour/stop9.html

Adger, N. (2000). Social and ecological resilience: are they related? *Progress in Human Geography, 24*(3), 347–364. https://doi.org/10.1191/030913200701540465

Ahn, J., Guarnieri, F., & Furuta, K. (2017). *Resilience: A New Paradigm of Nuclear Safety. From Accident Mitigation to Resilient Society Facing Extreme Situations. Resilience: A New Paradigm of Nuclear Safety. From Accident Mitigation to Resilient Society Facing Extreme Situations.* Springer. https://doi.org/10.1007/978-3-319-58768-4

Alinsky, S. (1989). *Rules for Radicals.* Vintage; a edition.

Alquist-Priolo Earthquake Fault Zoning (AP) Act. (2012). Retrieved from http://www.conservation.ca.gov/cgs/rghm/ap/Pages/index.aspx

Amir, S., & Kant, V. (2018). Sociotechnical Resilience: A Preliminary Concept. *Risk Analysis*, 38(1): 8–16.
Bennett, J. (2004). The Agency of Assemblages and the North American Blackout. *Public Culture*, *17*(3), 445–466. Retrieved from uake.usgs.gov/research/pager/prodandref/WaldEtAlECEESDYFI.pdf
Blume, J. A. (n.d.). John A. Blume—trailblazing earthquake engineer. Retrieved from https://engineering.stanford.edu/about/heroes/john-blume
Bonanno, G. (2004). Loss, Trauma, and Human Resilience. Have we underestimated the Human Capacity to Thrive After Extremely Aversive Events? *American Psychologist*, *59*(1), 20–28.
Bossu, R., Gilles, S., Mazet-Roux, G., Roussel, F., Frobert, L., & Kamb, L. (2011). Flash sourcing, or rapid detection and characterization of earthquake effects through website traffic analysis. *Annals of Geophysics*, *54*(6), 716–727. https://doi.org/10.4401/ag-5265
Boudia, S., & Jas, N. (2007). Risk and Risk society in Historical Perspective. *History and Technology*, *4*, 317–331.
Bourque, L., & Russell, L. (1994). *Experiences During and Responses to the Loma Prieta Earthquake.*
Brechin, G. (2006). *Imperial San Francisco, Urban Power, Earthly Ruin.* University of California Press.
Burby, R., & May, P. (1998). Making building codes an effective tool for earthquake hazard mitigation. *College of Urban and Public Affairs (CUPA) – Working Paper, 1991–2000, Paper 3.* Retrieved from http://scholarworks.uno.edu/cgi/viewcontent.cgi?article=1002&context=cupa_wp
California Seismic Safety Commission. (2001). *Findings and Recommendations on Hospital Seismic Safety.*
Chakos, A. (2006). Message to Locals in California Disasters: YOYO-You're on Your Own.
Chakos, A., Schulz, P., & Tobin, T. (2002). Making it Work in Berkeley: investing in Community Sustainability. *Natural Hazard Review*, 1.
Coen, D. (2013). *The Earthquake Observers- Disaster Science from Lisbon to Richter.* The University of Chicago Press.
Comfort, L. K. (2006). City at Risk: Hurricane Katrina and the Drowning of New Orleans. *Urban Affairs Review*, *XX*(X), 1–16. https://doi.org/10.1177/1078087405284881
Comfort, L. K., Boin, A., & Demchak, Chris C. (2010). *Designing Resilience: Preparing for Extreme Events.* University of Pittsburgh Press; 1 edition.
Contributors, W. (n.d.-a). "Bruce Bolt." In *Wikipedia, The Free Encyclopedia.* Retrieved from https://en.wikipedia.org/wiki/Bruce_Bolt
Contributors, W. (n.d.-b). "Henry J. Degenkolb." In *Wikipedia, The Free Encyclopedia.* Retrieved from https://en.wikipedia.org/wiki/Henry_J._Degenkolb

Contributors, W. (n.d.-c). "Unreinforced Masonry Building." In *Wikipedia, The Free Encyclopedia*. Retrieved from http://en.wikipedia.org/wiki/Unreinforced_masonry_building

Davis, M. (1998). *Ecology of Fear. Los Angeles and the Imagination of Disasters.* Vintage Books.

Folke, C. (2006). Resilience: The emergence of a perspective for social-ecological systems analyses. *Global Environmental Change, 16*(3), 253–267. https://doi.org/10.1016/j.gloenvcha.2006.04.002

Frickel, S., & Bess, V. (2007). Katrina, Contamination, and the Unintended Organization of Ignorance. *Technology in Society, 29*, 181–188.

Geschwind, C.-H. (2001). *California Earthquake, Science, Risk and the politic of Hazard Mitigation*. The Johns Hopkins university Press.

Gomart, E., & Hennion, A. (1999). A sociology of attachment: music amateurs, drug users. *The Editorial Board of The Sociological Review*. https://doi.org/10.1111/j.1467-954X.1999.tb03490.x

Gorney, C. (2009). Remebering the Surgery Inside the Freeway Collapse. Retrieved from http://oaklandnorth.net/2009/10/17/remembering-the-surgery-inside-the-freeway-collapse/

Grossi, P., & Muir-Wood, R. (2006). *The 1906 San Francisco Earthquake and Fire: Perspectives on a Modern Super Cat.*

Guarnieri, F. (2017). The Fukushima Daiichi Nuclear Accident: Entering into Resilience Faced with an Extreme Situation. In J. Ahn, F. Guarnieri, & K. Furuta (Eds.), *Resilience: A New Paradigm of Nuclear Safety. From Accident Mitigation to Resilient Society Facing Extreme Situations.* (pp. 1–17). Springer. https://doi.org/10.1007/978-3-319-58768-4

Hall, P., & Lamont, M. (2013). Social resilience in the neoliberal era. Cambridge University Press.

Hoffman, S. M., & Oliver-Smith, A. (2002). *Catastrophe and Culture. The Anthropology of Disasters.* (S. M. Hoffman & A. Oliver-Smith, Eds.). School of American Research Press, James Curry.

Inventory of the Joint Committee on Seismic Safety Records. (n.d.). Retrieved January 12, 2017, from http://www.oac.cdlib.org/findaid/ark:/13030/tf0290005q/

Jasanoff, S. (2014). Genealogies of STS. *Social Studies of Science, 42*(3), 435–441. https://doi.org/10.1177/0306312712440174

Johnson, L. (2011). *Insuring Climate Change: Science, fear, and value in reinsurance markets.* University of California Berkeley.

Juraku, K. (2015). Why Is It So Difficult to Learn from Accidents? In *International Workshop on Nuclear Safety: From Accident Mitigation to Resilient Society Facing Extreme Situations.* Berkeley.

Knorr-Cetina, K. (1999). *Epistemic Cultures: How the Sciences Make Knowledge.* Harvard University Press. Retrieved from https://books.google.com/books?id=WFEeib0Q9L0C&pgis=1

Knowles, S. (2011). *The Disaster Experts: Mastering Risk in Modern America*. The University of Pennsylvania Press.
Lagorio, H. J. (n.d.). Henry J. Lagorio.
Lane, S., Odoni, N., Landstrom, C., Whatmore, S., Ward, N., & Bardley, S. (2010). Doing flood risk science differently: an experiment in radical scientific method. *Transaction of the Institute of British Geographers*, *36*, 15–36.
Lash, S., Szerszynski, B., & Wynne, B. (1996). *Risk Environment and Modernity. Toward a New Ecology*. (L, Ed.) (Published). SAGE Publications Ltd.
Latour, B. (2005). What Is Given in Experience? *Boundary 2*, *32*(1), 223–237. https://doi.org/10.1215/01903659-32-1-223
Latour, B., & Girard Stark, M. (1999). Factures/Fractures: From the Concept of Network to the Concept of Attachment. *RES: Anthropology and Aesthetics*, *36*(Autumn), 20–31. Retrieved from http://www.bruno-latour.fr/sites/default/files/downloads/76-FAKTURA-GB.pdf
Law, J. (1990). Introduction: monsters, machines and sociotechnical relations. *The SociologicalReview*,38(1S),1–23.https://doi.org/10.1111/j.1467-954X.1990.tb03346.x
Lin, R.-G. I. (2017, May 3). Earthquake early warning system nets $10.2 million in Congress' budget deal. *Los Angeles Time*. Retrieved from http://www.latimes.com/local/lanow/la-me-ln-earthquake-early-warning-20170502-story.html
Lussault, M. (2007). *L'homme spatial, la construction sociale de l'espace humain*. Editions du Seuil.
Mahony, M., & Hulme, M. (2016). Epistemic geographies of climate change. *Progress in Human Geography*, 30913251668148. https://doi.org/10.1177/0309132516681485
Mazel-Cabasse, C. (2017). Hybrid Disasters—Hybrid Knowledge. In J. Ahn, F. Guarnieri, & K. Furuta (Eds.), *Resilience: A New Paradigm of Nuclear Safety. From Accident Mitigation to Resilient Society Facing Extreme Situations*. (pp. 337–351). Springer. https://doi.org/10.1007/978-3-319-58768-4
Myers, D., & Wee, D. F. (2005). *Disaster, Mental Health Services: A Primer for Practitioners*. Brunner-Routledge.
Nigg, J. M., & Mileti, D. S. (1998). *The Loma Prieta California Earthquake of October 17, 1989—Recovery, Mitigation and Reconstruction* (USGS Professional Paper No. 1553–D).
November, V., & Leanza, Y. (2015). *Risk, Disaster and Crisis Reduction*. Cham: Springer International Publishing. https://doi.org/10.1007/978-3-319-08542-5
Olson, R. A. (n.d.). Robert A. Olson. Retrieved from http://www.olsonassoc.com/utility/about.html
Perkins, J. B., Chakos, A., Olson, R. A., Tobin, L. T., & Turner, F. (2006). A Retrospective on the 1906 Earthquake's Impact on Bay Area and California

Public Policy. *Earthquake Spectra*, 22(S2), S237. https://doi.org/10.1193/1.2181527

Perrow, C. (2006). The Disaster after 9/11: The Department of Homeland Security and the Intelligence Reorganization. *Homeland Security Affairs*, *II*(Article 3). Retrieved from https://www.hsaj.org/articles/174%0A%0A%0A%0A

Quarantelli, E. L. (1998). *What is a Disaster?: A Dozen Perspectives on the Question*. Routledge.

Reghezza-Zitt, M., & Rufat, S. (2015). *Resilience Imperative: Uncertainty, Risks and Disasters*. (ISTE Press—Elsevier, Ed.).

Reghezza-Zitt, M., Rufat, S., & Djament-Tran, G. (2012). What Resilience Is Not: Uses and Abuses. Retrieved from http://cybergeo.revues.org/25554

Solnit, R. (2009). *A Paradise Built in Hell: The Extraordinary Communities that Arise in Disaster*. Viking.

Steinbrugge, K. (1968). *Earthquake hazard in the San Francisco Bay Area: a Continuing Problem in Public olicy*. Institute of Governmental Studies, University of California.

Steinbrugge, K. V. (n.d.). Karl V. Steinbrugge.

Strauss, J. A., & Allen, R. M. (2016). Benefits and Costs of Earthquake Early Warning. *Seismological Research Letters*, *87*(3), 765–772. https://doi.org/10.1785/0220150149

Taira, T., Silver, P. G., Niu, F., & Nadeau, R. M. (2009). Remote triggering of fault-strength changes on the San Andreas fault at Parkfield. *Nature*, *461*(7264), 636–639. https://doi.org/10.1038/nature08395

Tierney, K. J. (2001). *How will Social Science Help us Deal with Earthquake?* University of Chicago Press.

Timmerman, P. (1981). Vulnerability, Resilience and the Collapse of Society: A Review of Models and Possible Climatic Applications. *Environmental Monograph*. https://doi.org/10.1002/joc.3370010412

Tobriner, S. (2006). *Bracing for Disaster. Earthquake-Resistant Architecture and Engineering in San Francisco, 1838–1933*. The Bancroft Library, University of California Berkeley.

Tubbesing, S., & Mileti, D. S. (1994). *The Loma Prieta, California, Earthquake of October 17, 1989—Loss Estimation and Procedure*. Washington. Retrieved from http://books.google.com/books?hl=en&lr=&id=_E3wAAAAMAAJ&oi=fnd&pg=SL4-PA3&dq=Conducting+earthquake+preparedness+campaigns:+A+marketing+approach+%28BAREPP%29+&ots=IsskiLJyN6&sig=xbcbigVAp-ftRZaI0A-wseTWhks#v=onepage&q&f=false

Turner, F. (2004). Seventy Years of the Riley Act and Its Effect on California's Building Stock. In *13th World Conference on Earthquake Engineering*. Vancouver.

Walker, R. (1990). The PlayGround for US Capitalism? The Political Economy of San Francisco Bay Area in the 80's. In M. Davis, S. Hiatt, M. Kennedy,

S. Ruddick, & M. Sprinker (Eds.), *Fire in Hearth, The Radical Politics of Place in America*. Verso.

Weichselgartner, J., & Kelman, I. (2015). Geographies of resilience: Challenges and opportunities of a descriptive concept. *Progress in Human Geography, 39*(3), 249–267. https://doi.org/10.1177/0309132513518834

White, G. (1945). Human adjustment to floods – A geographical approach to the flood problem in the United-States. The University of Chicago. Retrieved from http://www.colorado.edu/hazards/gfw/images/Human_Adj_Floods.pdf

Wilshire, H. (n.d.). Collapsed Cypress Street Viaduct in Oakland, from the 1989 Loma Prieta earthquake. Retrieved from https://commons.wikimedia.org/wiki/File:022srUSGSCyprusVia.jpg

Wisner, B., Blaikie, P., Cannon, T., & Davis, I. (2006). *At Risk*. Routledge.

CHAPTER 3

Sociotechnical Resilience: From Recovery to Adaptation and Beyond, the Journey So Far...

Stephen Healy

Introduction

There remains a lack of clarity regarding the essential character of resilience, inconsistent with the increasing political and broader decision-making attention it attracts. In an earlier publication, a co-author and I discussed this "pervasive ambiguity" noting how resilience is characterized "as an outcome or process characteristic, as intrinsic or contingent, a matter of system structure or function, as a paradigm or as 'just an expression'" (Healy and Mesman 2014: 196). Much of the current literature reflects this imprecision tending, for the most part, to emphasize either "engineering resilience" (Holling 1996: 1473), understood as an ability to "bounce back" to recover a system state or condition equivalent to that existing prior to disturbance, or for a system to have the ability to recover from disturbance such that an essential function, or functions, are maintained (e.g. Meerow et al. 2016).[1] However, in addition to this lack of

S. Healy (✉)
University of New South Wales, Sydney, NSW, Australia
e-mail: s.healy@unsw.edu.au

S. Amir (ed.), *The Sociotechnical Constitution of Resilience*,
https://doi.org/10.1007/978-981-10-8509-3_3

clarity over the essential meaning of resilience, critical matters of power, control, and social structure are also often overlooked (e.g. Meerow et al. 2016). Smith and Stirling (2010) illuminate the grounds for such ambiguity and imprecision and articulate a requirement to "reflect on what precisely it is that is being made resilient, in the face of which specific dynamics, for whom and by what criteria this is good or bad, and whether such resilience is consequently problematic or not" (2010: 11). However, they do this in an ecologically focused journal. For this journal, resilience is simply an ecological imperative, which considerably frames the interpretation that they give to the challenges Smith and Stirling articulate. This journal, the mouthpiece of the Resilience Alliance, affords a significant platform for the articulation of questions regarding the relationship between resilience, understood ecologically, and social systems, a relationship that the Resilience Alliance articulates as social-ecological systems.[2] While this journal's perspective is marked by limited input from the social sciences, humanities and broader social theory,[3] it currently provides a significant source of information regarding relationships between resilience, understood ecologically, and human society. This is consequential because the role of societal dynamics in resilience is conspicuously absent from the broader resilience literature. Further, although socially informed approaches commonly reflect a democratic ethos and recommend that people, as members of an ensemble to be rendered resilient, be granted a meaningful role in arbitrating the questions that Smith and Stirling (2010: 11) articulate, unintended consequences may result from this. The outcome of such a democratic process might, for example, not only fail to endorse the ecological priorities emphasized by the Resilience Alliance but be understood, through the focus upon individual accountability implicit in a democratic process, to echo the neoliberal tendency to devolve responsibility to affected populations.[4] This exemplifies the problematic, frequently contested, but far from rare or inconsequential tensions commonly revealed by substantive engagement with the sociotechnical constitution of resilience (Healy and Mesman 2014).

A review of the recent literature was conducted to help inform the way this chapter interprets how resilience might best be conceived, managed, and acted upon and such tensions effectively addressed. The insights generated are applied to a case study focused upon the challenge that the rapid emergence of substantive demand-side energy provision is providing the Australian electricity industry. Australia currently has the world's highest installation rate for domestic rooftop solar photovoltaic panels (PV), for a range of reasons including policy support and high incident solar radiation levels (The Conversation 2016). Indeed, commentators have noted that "Australian consumers can already install significant amounts of rooftop solar and battery storage at a cost that is cheaper than electricity from the grid, and the uptake of these two technologies is likely to be 'unstoppable'" (Parkinson 2015a). This unheralded growth in installed PV has been marked by "double the rate of take-up (15 per cent of households on average) compared to the next country, Belgium where about 7.5 per cent of households have solar" (Vorrath 2016).

While a positive development for many, this is not how mainstream energy interests have received it. Although the resilience of energy and water systems are significantly enhanced by substantive domestic self-provision (Arcari et al. 2011),[5] a matter particularly well established in the case of distributed PV (NREL 2014), the Australian electricity industry has, for the most part, been unsupportive. In particular there has been a distinct "trend among some electricity networks to penalize or discourage the uptake of rooftop solar by imposing fixed tariffs or additional fees ... extending to battery storage, with one network accused of trying to lift charges to households with storage even though they are reducing peak demand" (Parkinson 2015b).

Hence, despite widespread recognition that solar PV enhances the sustainability and resilience of Australia's electricity industry and has benefits, to both communities and energy markets (Coleman and Teixeira 2016), incumbent electricity industry interests have been unsupportive.[6] While this evolving case study displays uncertainties common to all such dynamic developments, its character is directly pertinent to the current trajectory of electrical systems across both the developed and developing world.[7] It both illuminates the notion of an "energy transition,"[8] widely viewed as the key dynamic attaching to an effective jurisdictional response to climate change, and the substantial challenges to which "disruptive" technological change, identified as "such an improvement that it renders existing industries obsolete" (McConnell 2013), can give rise.[9] The conclusion of this chapter that "political will," and its translation into effective governance, are critical to, not only such "energy transitions," but also the effective enactment of sociotechnical resilience, more generally resonates with recent policy commentary (Hampton et al. 2017) and analysis. Amir and Kant discern, for example, "intentional and organized changes as the defining feature of sociotechnical resilience" (Amir and Kant 2018).

The first section, following, briefly scrutinizes recent interpretations of resilience and finds that the most widely found systemic interpretations are, typically, insensitive to the situated particulars required by effective analysis. An assemblage interpretation, which is a particularly powerful reading of Amir and Kant's (2017) analysis of sociotechnical resilience in terms of "mutually entangled" *hybrids* that are "both social and technical at the same time," is found to remedy this deficit. This interpretation focuses upon the collections of people and things constitutive of circumstances, and an ensemble understanding of these is found to best illuminate the situated dynamics from which resilience emerges. The following sections examine current tensions in the Australian electricity supply industry employing this insight. These tensions over changes in the Australian electricity industry are then further elucidated through the lens of transition management, an approach to fundamental structural change pioneered in the Netherlands for managing the emergence of both sustainability and the low carbon energy system that sustainability necessitates. Transition management powerfully illuminates how governance for resilience is critical for the achievement of future resilience, elements of which are examined in the final sections of the chapter.

Systemic or Situated?

Explanations of resilience, endeavoring to impart a comprehensive perspective, are commonly founded upon the major systems of interest, most typically framed by the binary distinctions central to the Western intellectual tradition (i.e., nature/culture, body/mind, fact/value, etc.), mapping the empirically evident relationships between them. Such a depiction might, for example, start with a system of government, or governance, and thence various dependent social and/or infrastructural systems, followed by the natural/ecological systems of interest, with the latter forming a "foundation" for the overall representation.

These systemic renditions echo traditional analytic frameworks and their grounding in the universalistic, totalizing tendencies of Western science. However, and arguably as a result, they commonly confound effective analysis by poorly engaging the specifics of locality, affected populations or situated, local processes. More recent thinking, deriving from contemporary continental thought, notably Deleuze, and recent socio-material theory,[10] reverses the emphasis on the universal, as against the situated, through emphasizing the notions of assemblage, or agencement,[11] understood as a dynamic arrangement of people and things focused through particular processes.[12] An assemblage, or agencement, entails not only a mapping of the relationships by which they are constituted but also the processes of emergence they might facilitate (Phillips 2006), in whatever domain, such as particular places or practices, they are applied to or found within. I use the term ensembles for these collections because, not only is "[t]his idea of a socio-material and sociotechnical ensemble … the most literal meaning of assemblage" (Farías 2010), but myself (Healy 2004) and others (e.g., Bijker 1995) have used this term to better portray the meaning-laden dynamics of these active, dynamic amalgamations.

Thus, although systemic characterizations reveal the paradigmatic universal quality so esteemed by Western science, ensembles amalgamate technical features, specific to particular places, resources and/or practices, with the social and moral meanings, and agential forces, that participants complementarily deploy to engender eventful circumstances. A key difference between the two being that situated, contextual particulars do not claim a universal status but rather conform, simply, to specific circumstances. Ensembles can be identified correlating to a particular practice or resource, such as water, energy, or automobility, or belonging to a specific place or particular community. The ensemble depiction, thus, facilitates not only an illumination of the situated specifics pertaining to a matter of interest but also the ways affected populations experience, understand and participate in them. The resultant governmental implications have explained how "the logic of systems is replaced with the contingency of assemblages to reveal how pluralism, not elitism, can produce more ambitious and politicized visions of the future" (Gillard et al. 2016). These matters are further examined in the final sections that follow.

Energy, Energy Everywhere

In the industrialized world, electricity systems evolved to take a centralized form in which large-scale electricity generation stations, primarily fossil-fueled, distributed the generated power to end-users via a transmission "grid." This applied the then available technological options focused by an "economics of scale," which made for a compelling logic in the late nineteenth and early twentieth centuries. Although still incumbent in many jurisdictions, recent decades have seen this centralized logic challenged by cost-effective decentralized, point-of-use, generation options, rendered available by the emergence of distributed renewable generation technologies whose prices continue to fall.[13] Australia has witnessed the massive growth in domestic rooftop solar PV described earlier as part of this transformation. Recently, the resultant proliferation of "prosumers"[14] has been accelerated by the emergence of new energy storage technologies in the domestic market.[15] Tesla, for example, has specifically targeted the Australian domestic market with its new Powerwall Battery technology (ABC 2016).[16] However, although "distributed PV can significantly increase the resiliency of the electricity system" (NREL 2014), and despite widespread recognition of the problems facing Australia's electricity industry, it has been unsupportive (e.g. Parkinson 2015b). So although "Australia's National Energy Market is hamstrung by an outdated, 'dumb' grid, and must be updated to face the realities of low carbon, low marginal cost energy generation" (Quinn 2016), the industry has remained moored by the logic of yesteryear. This has been particularly counterproductive with "the lingering conservatism of… market operators and policy makers and regulators, who over the last five years all but ignored new technologies and stuck to their projections of increased demand in justifying huge spending on network infrastructure" (Parkinson 2016a).

The economically punitive reactions to the rise of domestic solar PV described above (Parkinson 2015b) bear witness to these arguments. So although "Australia's electricity regulatory frameworks … require rethinking and adaption" to accommodate current challenges (Clean Energy Council 2015: 2) and the "critical need of reform" (Wood 2016), this indicates may be on the horizon (Parkinson 2016d), little change is currently evident.[17] Complicating this is that "[a]s numerous people have mentioned, including the head of China's State Grid, transitioning … [an] … energy system is not so much a technology issue, as a cultural and political issue" (Parkinson 2016b), a matter underlined by Thomas Hughes, the leading historian of electricity systems.[18] So, although "the transition from the outdated dirty power system…to the smart, flexible and cleaner power system of the future" (Parkinson 2016b) has many key technical dimensions, this is also, critically, "a cultural and political issue." While the technological aspects of this "transition" are customarily considered unchallenging, requiring little beyond the deployment of available technologies,[19] the business and regulatory challenges are proving to be significantly greater hurdles. However, the broader lifestyle and cultural aspects of this "transition" have been given little informed attention to date. Expert commentators note that "designing electricity markets for the prosumer era could maximize residential and commercial energy efficiency efforts, democratize demand-response and prepare society for ubiquitous distributed clean energy technologies." They do, however, add that "this can be achieved only if proponents are able to recognize and support prosumer markets differentiated by services, role and function, and anticipate a series of compelling caveats and complexities" (Parag and Sovacool 2016).

We should, perhaps, thus be unsurprised that the response of Australian government and regulatory authorities to the current changes to Australian energy markets remains hesitant, tenuous and out of touch with the speed and disruptive quality of these changes.[20] This is no better illustrated than by the "Electricity Network Transformation Roadmap" project, a partnership between CSIRO, Australia's national science agency, and the Energy Networks Association (ENA), the peak body of Australian energy transmission and distribution businesses, which has a timeframe of 2017–2027. So although "[a]cross the eastern seaboard served by the National Electricity Market (NEM) demand is collapsing and heading towards territory not seen since the last millennium.........The reduction in demand has clearly blindsided both industry and government, which continue to operate as though demand growth must inevitably return" (Sandiford 2014). As a result, the rapid regulatory attention this demands remains absent and those in a position to deliver this remain confident that it can wait a decade.[21]

Smart Energy Futures (or Not?)

Although many "prosumers" have responded to the "trend among some electricity networks to penalize or discourage the uptake of rooftop solar by imposing fixed tariffs or additional fees" (Parkinson 2015b) by leaving the grid, much of the traditional electricity industry appears unperturbed by these changes. This has encouraged the identification of a "death spiral" for the traditional electricity industry (Sandiford 2014),[22] although many view the grid as a pivotal feature of a future, sustainable electricity system.[23] This would, however, be a very different system to the current one. Demand is currently matched to available supply for five-minute intervals for which generators bid,[24] rendering the maintenance of power quality, primarily the uniform character of voltage and supply frequency, unchallenging. Embedded renewable generation, from intermittent generating sources, significantly increases both the difficulty of effectively matching supply to demand and of maintaining power quality. Future grids will likely necessitate extensive real-time monitoring and computational resources to cope with these challenges. While there are a number of such "smart grids" currently in operation there is limited experience in scaling them to a jurisdictional scale. One scenario contemplated for such scales is of using a number of interconnected "smart grids" to constitute a future scaled-up jurisdictional scale system. The least challenging aspects of these systems are, generally considered, to be technological although the economic[25] and regulatory challenges these pose remain poorly addressed, while the "forms of life" they might inform have been little considered to date.[26] While a "smart grid" matched to these developments would be cost effective and improve resilience: "[t]he traditional market arrangement is already broken" and driving people off the grid (Stewart 2016); institutional preparations for this "transition" are underwhelming, and "prosumers" have been vociferously articulating their dissatisfaction (Solar Citizens 2016a, b).

So against both popular sentiment and insights such as "[i]ntegrating centralized and distributed system models may maximize advantages and minimize limitations .... An integrated system comprising linked provisional infrastructure at multiple scales may offer the best way to build resilience at all levels – from resource producer to resource user" (Arcari et al. 2011: 5); Australia remains robustly in the grip of historical logics. This is has been widely observed with commentators noting, "[i]nstitutional inertia is a major issue" (Parkinson 2016f). A notably similar "inertia" has previously been observed of the Australian institutional capacity for change (Matthews 2011). Matthews noted, of the Australian institutional approach to climate adaptation (2011: 14–15), that "in spite of an emergent institutional capacity… there appears little willingness to view it as an immediate institutional imperative that compels policy or planning change," an observation particularly prescient for the case discussed here. Indeed, a dearth of coordination of climate and energy policy has been directly connected with Australia's "constrained" sustainable energy transition (Warren et al. 2016). While Australia certainly has an "emergent institutional capacity" to manage resilience[27] this is meticulously conventional (i.e. being, primarily, grouped into areas such as "disaster resilience" and "organizational resilience") and distinguished by limited integrative capacity. So, although the established wisdom is that planning for resilience must be long term, comprehensive, and coherent over time, this has been notable by its absence in Australia. Policy and business approaches have, rather, been fragmented and underpinned by emphases on business as usual, with the many downsides attendant upon such limited and partial perspectives.[28] This is no better witnessed than by the presence of climate skeptics within the ranks of the current Australian government.[29] So, analogous to the Resilience Alliance's very partial perspective on resilience, the Australian government has a, correspondingly, limited capacity to determine "what precisely it is that is being made resilient, in the face of which specific dynamics, for whom and by what criteria this is good or bad, and whether such resilience is consequently problematic or not" (Smith and Stirling 2010: 11). The current disjointed and fragmented Australian approach to the energy transition, and predisposition to favor status quo interests and perspectives (Warren et al. 2016) should, thus be unsurprising. A particularly coherent, and among the best-known, way of managing a transition, that of Transition Management, with which the Netherlands has jurisdictional experience and is used internationally for the analysis of sustainability and energy transitions,[30] is outlined in the following section.

Transition Management for Resilience

Whereas the Australian institutional encounter with the energy "transition," detailed above, has encountered significant resistance from incumbent institutions the Netherlands has been more welcoming to analogous developments and pioneered a formal governmental approach to "transitions management" centered on the multi-level perspective (MLP). Focused upon achieving sustainability,[31] the MLP involves the identification of three levels focused through an entrenched "sociotechnical regime," and bounded by "landscape developments" above and "technological niches" below. In the MLP, five dimensions are identified to represent the trajectory of an entrenched sociotechnical regime, such as that of automobility or, until recently, the hegemony of fossil fuel technologies in the energy domain. These five dimensions are those of: science; technology; policy; culture; and markets, user preferences. This sociotechnical trajectory is bounded by landscape developments above and, occasionally, challenged by the emergence of developing "technological niches" below. Landscape developments are intended to encompass the legal, regulatory and economic influences on a "socio-technical regime," including that of government.

Technological niches develop as emergent technologies grow, gain influence and dynamism and, thus, may eventually challenge the predominant sociotechnical regime. For example, in the case study described here, domestic rooftop solar is growing and presenting a challenge to the predominant Australian fossil fuel regime. It is important to highlight that the MLP emphasis on regime change is not consistent with those on continuity, recovery and adaptation at the core of, particularly ecological systems, resilience theory. The links between these two, apparently opposed, theories are, however, significant (Stockholm Environment Centre 2010) and considered "to provide a bridging opportunity to share lessons concerning the governance of both" (Smith and Stirling 2008: 2), although the differences between them are profound. Smith and Stirling (2008: 11) summarize that "Adaptive management[32] is more concerned with resilience that maintains social-ecological system functions and avoids large-scale collapse; whilst transition management is concerned with transformation to a sustainable socio-technical system over the longer-term."

It is, however, clear that these competing emphases on continuity and change are both critical to the achievement and maintenance of long-term global sustainability, and so extensive attention has been directed to the linkages between them. Those that have done this (e.g. Smith and Stirling 2008; Stockholm Environment Centre 2010; Meadowcroft 2011: 545–547) stress "the involvement of societal stakeholders, and ultimately … the approbation of political authorities" (Meadowcroft 2011: 547) as pivotal and it is to such matters of governance that this account turns below. A sense of the aims of shared governance can be gained from Smith and Stirling who note that "transition management injects goal-directing processes into socio-technical transformations. There are multiple governance challenges: collectively envisioning viable sustainability goals; nurturing promising niches; building supportive constituencies of actors, institutions and markets; and continually anticipating, learning and adapting" (2008: 9), with the latter echoing a key feature of adaptive management.

Governance for Resilience

The MLP facilitates insights into how the "niche" politics described in this chapter may develop with it widely recognized that "the most important challenge for energy transition governance lies in ensuring that future governments will remain credibly committed to overall transition visions and goals" (Laes et al. 2014: 1143). This, however, must be achieved within broader, and widely acknowledged, constraints such as "[h]ow do challenging bottom-up governance initiatives confront the deeply structural forms of economic power vested in current global patterns of system reproduction? How are different bodies of knowledge and interests in social-ecological systems negotiated? How is consent achieved, and how is dissent reconciled?" (Smith and Stirling 2010). Recent work in this tradition has shown how niche growth can be either contained or facilitated (e.g. Hess 2016: 42). Such analyses generally assume an existing normative foundation for a transition that informs "managerialist steering and consensus building" (Gillard et al. 2016: 260). However, as the case of Australia well illustrates, conventional managerialism can be unhelpful and, arguably, damaging. A recent study of climate change governance alert to such problems suggests that "focusing on the contingent relations between various actors (human and nonhuman) and their assemblages (e.g. an industry or a community) instantly opens up possibilities for more radical innovation and adaptability beyond the discursive confines of a functionalist system perspective" (Gillard et al. 2016: 260). In addition, although a focus on contingency and the complexities attaching to assemblages can generate profound uncertainties "unlike conventionally linear approaches to governance, which focus on realizing clearly defined goals and which therefore seek to minimize ambivalence, strategies of sustainable transition management are frequently informed by more reflexive modes of governance and by a willingness to embrace certain forms of ambivalence" (Walker and Shove 2007: 222). Walker and Shove make the point that "[t]here is a politics to the governance of transitions that works with and contributes both to the ambivalence of sustainability as a discursive category and to the playing out of power in two key arenas, in the definition of the 'system' in question, and in specifying modes and moments of intervention" (Walker and Shove 2007: 222). They further argue that ambivalence is "the very stuff of a dynamic, critical and questioning liberal politics" (2007: 223) and, therefore, that the "critical political challenge is to design forms of governance that foster and sustain ambivalence" (2007: 223).

To this end, they advocate fostering "grey zones and interstices within existing orders" such that "positively generative structures of ambivalence come into being" (2007: 223). Such "governance on the inside" has previously been described as an "ideal-typical" approach to the governance of sociotechnical systems (Smith and Stirling 2007).[33] Those that have broached such complexity recommend "focusing on the contingent relations between various actors (human and nonhuman) and their assemblages (e.g., an industry or a community)" and "the interpretive and strategic actions of influential actors before, during, and after moments of crisis and agitation" in order to foster dialogue regarding "whose vision of a climate compatible future is being pursued and along which pathways?" (Gillard et al. 2016: 260–1). In other words, the suggestion is that twenty-first century democracies require fundamental reform to facilitate not only broad-based debate and deliberation but also both institutional and societal reflexivity.

Conclusion

Resilience, particularly the constitution of sociotechnical resilience, reinstates, if this ever were not the case, our attention firmly back on matters of governance. As Meadowcroft (2011) above clarifies, "the challenges of the future" are far from undemanding in these regards. Analytically this necessitates that we open "up possibilities for more radical innovation and adaptability beyond the discursive confines of a functionalist system perspective" (Gillard et al. 2016: 260–1) through analysis of assemblages or, as argued here, ensembles, rather than more narrowly delineating sociotechnical systems. This is, however, more straightforward than the requirement to engage the reality of politics, agency and power in whatever context an ensemble of interest operates. This "reality," with a firm ensemble focus will, necessarily, be micro-, as against macro-, political requiring not the generic, neo-positivist insights of the political sciences but, rather, the situated insights of the (new) humanities.[34]

This is no better illustrated than by the case study of this chapter. This shows that "one task for policy is to not fall prey to special interests, hypes and undue criticisms...... All new energy technologies come with specific dangers and hazards, which have to be anticipated and addressed. For sustainable energy there are no technical fixes, nor are there perfect instruments. There is a need for policy to be more concerned with system change. The capacity to do so has to be created" (Kemp 2010: 311). In Australia currently structurally powerful political and economic actors are, successfully, thwarting an otherwise well-placed potential energy transition. However, the dynamism of many of the organizations facilitating renewable energy in Australia suggests a potential for fundamental change. Indeed one of the most insightful commentators on these matters recently noted, "[t]he change is upon us and it's all OK. We just need our regulators and our politicians to catch up" (Parkinson 2017a). For this to happen "politicians" and prosumers will have to agree on a normative basis for their actions. Such "diverse ecologies of participation" are clearly "important for moving towards more deliberately reflexive governance for sustainability and attending to the politics of socio-technical transitions, at least when it comes to participation and 'the public'" (Chilvers and Longhurst 2016: 603). However, the "reflexive modes of governance and ... willingness to embrace certain forms of ambivalence" (Walker and Shove 2007: 222) this may require highlight important future research foci. Meadowcroft underlines that there is "much more to be done to explore avenues for change in democratic polities," and suggests "[t]he point is ... that ... the state ..... will have to respond to new challenges, be recast in new institutional forms, and act more cooperatively with a complex array of social forces if it is to handle the challenges of the future" (Meadowcroft 2011: 549). He constructively suggests "public education, the building of reform coalitions and institutional innovation" (2011: 549) as potential foci for this, although these await future study.

A recent commentary on matters surrounding this chapter's case study noted that it "leaves Australia with ... a grid designed by idiots. This is not Grid 2.0" (Parkinson 2017b). Smith and Stirling, among the more seasoned analysts of such matters, observe that "a more politically-inclined perspective sees hope in the messiness and slipperiness of processes beyond the reach of the more managerial forms of transition management. Reflexive governance of a sort is already practiced on a day-to-day basis by the social groups and movements who lobby to get their social-ecological priorities heard by political authority and economic power, and who create alternative niches offering inspiring solutions for others to adopt and adapt" (2008: 21). So although I must conclude at a point where "Grid 2.0" remains illusory in the Australian context, "inspiring solutions" reflexively developed elsewhere illuminate promising alternatives. The implication from all this is that contemporary challenges necessitate challenging *realpolitik* is of no small consequence for not only energy transitions but also the successful management of sociotechnical resilience more generally.

Notes

1. While "engineering resilience" was the original ecological definition of resilience, more sophisticated ecological notions have been developed, most recently Panarchy (Gunderson and Holling 2002), which recognizes the importance of intra-systemic change, involving not only society but

also the many different scales and rates at which changes, and adaptations, occur. Some (e.g. Meerow et al. 2016) have, further, noted how resilience can act as a "boundary object" to which different audiences attribute different meanings!

2. Defined as "complex, integrated systems in which humans are part of nature" (www.resalliance.org/key-concepts accessed 16/3/16), reflecting the ecological priority attaching to their systems ecology viewpoint.
3. Although recently subject to some study (e.g. Stone-Jovicich 2015).
4. Neoliberalism celebrates the autonomous individual of economic theory, complementary to its derision for state power, by devolving as much of that power to individuals as it can, commonly without resources (Hamann 2009), thereby constructing resilience as an individual responsibility (Josepth 2013).
5. This study "sought to understand how community-scale energy and water systems influence community adaptive capacity and resilience to climate change" (2011: 5).
6. A new report by the Australian Energy Market Operator (AEMO), mid-2017, while this chapter was being finalized, suggests that positive changes, although many current political interests would oppose them, could be underway (Parkinson 2017c). This is due, in no small part, to AEMO's new CEO, appointed March 2017, who previously held the position of Chair of the New York State Public Service Commission (NYPSC). NYPSC is internationally recognized for recent innovative regulatory reforms lowering consumer energy costs while emphasising a more resilient and reliable power system, with a focus upon the facilitation of decentralized demand-side power supply options.
7. The status of these challenges in the developing world will not be further discussed in this chapter (but see, for example: Khoury et al. 2016 and www.wri.org/blog/2016/05/rise-urban-energy-prosumer). This is further attested to by both the African Union's "African Renewable Energy Initiative" (AREI) and France and India's launch of the "International Solar Alliance" at the UNFCCC's COP 21, December 2015.
8. This notion of a fundamental systemic change in energy production, and use, is typically considered via the perspective of 'transition management' discussed below.
9. See McConnell (2013) for a discussion of solar as a 'disruptive' technology.
10. Actor Network Theory is among more the recent manifestations of socio-material theory of specific relevance here, elaborated insightfully by major practitioners such as Bruno Latour and John Law. Muller (2015) is of specific relevance to this discussion.

11. Closely related to Foucault's *dispositif*, which places more emphasis upon power/knowledge (see: Braun 2014)
12. Dynamics subject to increasing scholarly attention (e.g. Chilvers and Longhurst 2015).
13. In much of the developing world, discussed in endnote 7 above, electricity systems are, only now, starting to be organized on the basis of the currently available cost effective distributed renewable generation technologies making for very different challenges (See: Healy et al. 2017 for further details).
14. An energy consumer who produces energy has come to be known by this term because they both *produce* and *consume* electricity.
15. This was a fairly revolutionary development because the intermittency of the new technologies, which require an ongoing supply of their "fuel" (i.e. the sunshine, wind etc.) to generate power, was effectively rendered unproblematic. Opponents of these changes commonly couch their arguments in terms of a need for "baseload" generation, which is ongoing because their fuel supply, such as coal, is not "intermittent." Today, such arguments have rapidly become an historical anachronism (Parkinson 2017d).
16. While among the best known, this Tesla product is only one of a number of new, primarily, lithium-ion batteries that have recently come onto the market (ABC 2015). The very rapid evolution of these technologies is shown by the launch of a new Powerwall 2.0 version of these batteries a little less than a year after their initial launch but at about half the price and with double the storage capacity of the initial version (Mountain 2016).
17. With developments over the months since this was originally written only underscoring this, although see endnote 6 above.
18. "Electric power systems embody the physical, intellectual, and symbolic resources of the society that constructs them. Therefore in explaining the changes in the configuration of power systems, the historian must examine the changing resources and aspirations of organizations, groups, and individuals...electric power systems, like so much other technology, are both causes and effects of social change" (Hughes 1983).
19. New prosumer markets have technical challenges beyond the mundane, requiring not only that distributed generation is matched to consumption but also that power quality is adequately maintained across a system, requiring extensive computational and monitoring resources. To date these challenges have been only been successfully met in limited domains although the prevalent technical attitude to them is sanguine (the author, who is a PhD Electrical Engineer, suspects that this confidence is misplaced).

20. While this chapter was drafted in 2016 revisions were made in 2017 and recent developments, detailed by Parkinson (2017a), underline this observation.
21. Parkinson (2016c) details some of the "cracks" beginning to show in the system as a result. At the time of writing, renewable energy technologies were enduring repeated, sustained media attacks (Parkinson 2016h).
22. In addition to the identification of a commercial dilemma for the industry, many commentators point out that as customers leave the grid and prices rise, which is occurring, only those least able to afford the new premiums remain utility customers. The commercial industry dilemma has been more broadly recognized and discussed (e.g. *The Economist* 2017).
23. In which it would take more of a storage, bidirectional role rather than acting simply as a one-way conduit from the sites of electricity production to the sites of electricity consumption.
24. A recent attempt to match this to the interval over which these are paid, currently 30 minutes, which would accelerate battery storage but disadvantage market incumbents, has been put on hold (Parkinson 2016g).
25. While expected to require particularly innovative tariff structures there has been limited experience in implementing these to date. Under current tariff structures some "prosumers" are even paying utilities for electricity that they, themselves, generate (Parkinson 2016e).
26. Although "forms of life" is a term coined by Wittgenstein, it is widely used in Science and Technology Studies to identify the commonly dynamic ways we might live with and through technologies, with, in this case, those associated with electricity particularly significant in shaping the character and content of everyday life.
27. See, for example, www.organisationalresilience.gov.au/Pages/default.aspx
28. Not only are the emissions intensive and, more broadly, environmentally damaging aspects of conventional industry, but also the socially divisive impacts of the privileging of some over the many.
29. Including ministers (www.theguardian.com/australia-news/2016/jul/19/matthew-canavan-says-there-is-uncertainty-around-cause-of-climate-change).
30. Having given rise to many international developments such as the journal *Environmental Innovation and Societal Transitions* (www.journals.elsevier.com/environmental-innovation-and-societal-transitions/).
31. Applied to Dutch energy policy at the end of the twentieth century (Rotmans et al. 2001).
32. Adaptive Management is a widely recognized environmental method of optimizing sustainability by iteratively adapting to changing conditions over time.

33. The other "ideal-typical" approach being "governance on the outside," which assumes an objectively knowable sociotechnical system (Smith and Stirling 2007).
34. The author's current institutional affiliation to Environmental Humanities is an exemplary example. One founding principle of this new discipline being the annulment of human exceptionalism, which is a central feature of the traditional humanities.

References

ABC (2015), Home battery storage to 'revolutionise' solar industry in Australia: Climate Council report, (accessed online 26/7/16: http://www.abc.net.au/news/2015-10-21/home-battery-storage-to-revolutionise-solar-industry/6870444).

ABC (2016), Explained: The Tesla Powerwall and what it means for Australia's energy market, (accessed online 19/7/16: http://www.abc.net.au/news/2016-02-02/tesla-powerwall-what-it-means-for-australia's-energy/7130392).

Amir, S. and Kant, V. (2017), Sociotechnical Resilience: A Preliminary Concept. *Risk Analysis*, 38: 8–16. https://doi.org/10.1111/risa.12816.

Arcari, P., Biggs, C., Maller, C., Strengers, Y., Horne, R. and Ryan, C. (2011), Resilient Urban Systems: a Socio-Technical Study of Community Scale Climate Change Adaptation Initiatives, Final Report for the Victorian Centre for Climate Change Adaptation Research (accessed 9/2/16 at: http://www.vcccar.org.au/sites/default/files/vcccar/rusfr/index.html#/1/).

Bijker, W. (1995), *Of Bicycles, Bakelites, and Bulbs: Toward a Theory of Sociotechnical Change*, MIT Press.

Braun, B.P. (2014), A new urban dispositif? Governing life in an age of climate change, *Environment and Planning D: Society and Space*, 32: 49–64.

Chilvers, J. and Longhurst, N. (2015), A Relational Co-Productionist Approach to Sociotechnical Transitions, Science, Society and Sustainability (3S) Research Group Working Paper, University of East Anglia (accessed 1/4/16: https://uea3s.files.wordpress.com/2015/12/chilvers-and-longhurst-3swp-2015-27.pdf).

Chilvers, J. and Longhurst, N. (2016), Participation in Transition(s): Reconceiving Public Engagements in Energy Transitions as Co-Produced, Emergent and Diverse, *Journal of Environmental Policy & Planning*, 18:5: 585–607.

Clean Energy Council (2015), Evaluation Methodology of the Value of Small Scale Embedded Generation and Storage to Networks, Task FPDI TA-2C for the Clean Energy Council, July 2015.

Coleman, K. and Teixeira, A. (2016), Why community-scale solar is solar electricity market's latest darling, *Renew Economy* 27 May 2016 (accessed 16/8/16:

http://Renew Economy.com.au/2016/community-solar-a-cost-effective-bridge-away-from-net-metering-43987).

Farías, I. (2010), "Introduction: decentring the object of urban studies" in *Urban Assemblages: How Actor-Network Theory Changes Urban Studies*, Farías, I. and Bender T (eds), Routledge (London and NY): 14.

Gillard, R., Gouldson, A., Paavola, J. and Van Alstine, J. (2016), Transformational Responses to Climate Change: Beyond a Systems Perspective of Social Change in Mitigation and Adaptation, Wiley *Interdisciplinary Reviews: Climate Change*, 7(2): 251–265 (accessed 30/8/16: http://onlinelibrary.wiley.com/doi/10.1002/wcc.384/full).

Gunderson, L and Holling C.S. (Eds.) 2002. *Panarchy: understanding transformations in human and natural systems.* Washington: Island Press.

Hamann, T. H. (2009), Neoliberalism, Governmentality and Ethics, *Foucault Studies*, 6: 37–59.

Hampton, S, de la Cruz, J. and Huenteler, H. (2017), Political Consensus and the Energy Transition, 'Issue Brief' Atlantic Council Leadership Program (accessed 27/5/17: http://www.atlanticcouncil.org/publications/issue-briefs/political-consensus-and-the-energy-transition).

Healy S.A. (2004), A 'Post-Foundational' Interpretation of Risk – Risk as 'Performance, *Journal of Risk Research* 7(3): 277–296.

Healy, S. and Mesman, J. (2014), "Resilience: Contingency, Complexity and Practice" in: *Vulnerability in Technological Cultures: New Directions in Research and Governance*, A. Hommels, W. Bijker and J. Mesman (eds), The MIT Press, 155–177.

Healy S, Munro, P.G. and van der Horst, G. (2017), Energy justice for all? Rethinking Sustainable Development Goal 7 Through Struggles over Traditional Energy Practices in Sierra Leone, *Energy Policy*, 105: 635–641.

Hess, D. (2016), The Politics of Niche-Regime Conflicts: Distributed Solar Energy in the United States, *Environmental Innovations and Societal Transitions*, 19: 42–50 (accessed 1/9/16: http://www.sciencedirect.com/science/article/pii/S2210422415300174).

Holling, C.S. (1996), "Engineering resilience versus ecological resilience" in: Shulze, P. (Ed.), *Engineering Within Ecological Constraints.* National Academy Press, Washington DC: 31–44.

Hughes, T.P. (1983), *Networks of Power: Electrification in Western Society, 1880–1930*, Johns Hopkins University Press.

Josepth, J. (2013), Resilience as Embedded Neoliberalism: a Governmentality Approach, *Resilience: International Policies, Practices and Discourses*, 1: 38–52.

Kemp, R. (2010), "The Dutch Energy Transition Approach," *Int Econ Econ Policy*, 7: 291–316.

Khoury, J., Mbayed, R., Salloum, G., Monmasson, E. and Guerrero, J. (2016), Review on the Integration of Photovoltaic Renewable Energy in Developing

Countries – Special Attention to the Lebanese Case, *Renewable and Sustainable Energy Reviews*, 57: 562–575.

Laes, E., Gorissen, L. and Nevens, F. (2014), A Comparison of Energy Transition Governance in Germany, The Netherlands and the United Kingdom, *Sustainability*, 6: 1129–1152.

Matthews, T. (2011), Operationalising Climate Adaptation through Institutional Change: Conceptual and Empirical Insights, presented at the 3rd World Planning Schools Congress, Perth (WA), 4–8 July 2011 (accessed 26/8/16: http://www98.griffith.edu.au/dspace/bitstream/handle/10072/44159/73608_1.pdf;jsessionid=35810D3364595D747E60631341386D81?sequence=1).

McConnell, B. (2013), Solar Energy: This Is What A Disruptive Technology Looks Like, http://www.resilience.org/ April 25, 2013 (accessed 30/8/2016: http://www.resilience.org/stories/2013-04-25/solar-energy-this-is-what-a-disruptive-technology-looks-like).

Meadowcroft, J. (2011), "Sustainable Development' in: *The SAGE Handbook of Governance*, Bevir, M. (ed.), Sage Publications: 535–549.

Meerow, S., Newell, J.P., and Stults, M. (2016), Defining Urban Resilience: A Review, *Landscape and Urban Planning*, 147: 38–49.

Mountain, B. (2016), Tesla's price shock: Solar + battery as cheap as grid power *Renew Economy* November 4, 2016 (accessed online 22/6/17: http://reneweconomy.com.au/teslas-price-shock-solar-battery-as-cheap-as-grid-power-22265/).

Muller, M. (2015), 'Assemblages and Actor-networks: Rethinking Socio-material Power, Politics and Space,' *Geography Compass* 9(1): 27–41.

NREL (National Renewable Energy Laboratory) (2014), Distributed Solar PV for Electricity System Resiliency: Policy and Regulatory Considerations, NREL/BR-6A20–62631, November 2014.

Parag Y. and Sovacool, B.K. (2016), 'Electricity Market Design for the Prosumer Era,' *Nature Energy*, 1: 1–6.

Parkinson, G. (2015a), Solar and battery storage already cheaper than grid power in Australia, *Renew Economy* 16 July 2015 (accessed online 19/7/16: http://RenewEconomy.com.au/2015/solar-and-battery-storage-already-cheaper-than-grid-power-in-australia-66169).

Parkinson, G. (2015b), 'Network charges may penalize uptake of battery storage, as well as PV,' *One Step off the Grid*, July 20, 2015 (accessed online 25/7/16: http://onestepoffthegrid.com.au/network-charges-may-penalise-uptake-of-battery-storage-as-well-as-pv/).

Parkinson, G. (2016a), 'Household solar and storage to change grid, says market operator,' *Renew Economy* 16 June 2016 (accessed online 26/7/16: http://RenewEconomy.com.au/2016/household-solar-and-storage-to-change-grid-says-market-operator-87182).

Parkinson, G. (2016b), Regulator opens up new battleground over Australia's dirty, dumb grid, *Renew Economy* 21 July 2016 (accessed online 26/7/16: http://RenewEconomy.com.au/2016/regulator-opens-new-battleground-australias-dirty-dumb-grid-31368).

Parkinson, G. (2016c), Australian energy markets have echoes of Enron crisis in California, *Renew Economy* 15 August 2016 (accessed online 17/8/16: http://RenewEconomy.com.au/2016/australian-energy-markets-echoes-enron-crisis-california-57123).

Parkinson, G. (2016d), 'Regulator to push networks to consider alternatives to poles and wires,' *Renew Economy* 14 July 2016 (accessed online 26/7/16: http://RenewEconomy.com.au/2016/regulator-push-networks-consider-alternatives-poles-wires-26589).

Parkinson, G. (2016e), Solar households now have to pay to use their own solar output, *Renew Economy* 31 March 2016 (accessed online 28/7/16: http://RenewEconomy.com.au/2016/solar-households-now-have-to-pay-to-use-their-own-solar-output-52794).

Parkinson, G. (2016f), Gas bubble looms as energy ministers baulk at zero emissions target, *Renew Economy* 26 August 2016 (accessed online 26/8/16: http://RenewEconomy.com.au/2016/regulator-delays-rule-change-that-could-accelerate-battery-storage-49231).

Parkinson, G. (2016g), Regulator delays rule change that could accelerate battery storage, *Renew Economy* 22 August 2016 (accessed online 23/8/16: http://RenewEconomy.com.au/2016/gas-bubble-looms-energy-ministers-baulk-zero-emissions-target-30072).

Parkinson, G. (2016h), How gas generators cashed in and exploited hot water load, *Renew Economy* 6 October 2016 (accessed online 12/10/16: http://Renew Economy.com.au/2016/how-gas-generators-cashed-in-and-exploited-hot-water-load-98349).

Parkinson, G. (2017a), The great divide over Australia's energy future, *Renew Economy* 22 May 2017 (accessed online 26/5/17: http://reneweconomy.com.au/coalitions-war-on-cheap-power-when-fools-design-energy-policy-57221/).

Parkinson, G. (2017b), Coalition's war on cheap power: When fools design energy policy, *Renew Economy* 19 June 2017 (accessed online 22/6/17: http://reneweconomy.com.au/the-great-divide-over-australias-energy-future-52410/).

Parkinson, G. (2017c), Consumers, PV and storage critical to low carbon grid: AEMO, *Renew Economy* 30 June 2017 (accessed online 3/7/17: http://reneweconomy.com.au/consumers-pv-and-storage-critical-to-low-carbon-grid-aemo-57582/).

Parkinson, G. (2017d), '"Baseload": An outdated term that should not be confused with "reliability",' *Renew Economy* 28 June 2017 (accessed online 6/7/17:

http://reneweconomy.com.au/baseload-an-outdated-term-that-should-not-be-confused-with-reliability-34961/).

Phillips, J. (2006), Agencement/Assemblage, *Theory, Culture and Society*, 23: 2–3.

Quinn, T. (2016), Time to ditch the "dumb" grid and embrace a smart energy future, *Renew Economy* 17 August 2016 (accessed 18/8/2016: http://Renew Economy.com.au/2016/time-to-ditch-the-dumb-grid-and-embrace-a-smart-energy-future-91220).

Rotmans, J., Kemp, R. and van Asselt, M. (2001), More evolution than revolution: transition management in public policy, *Foresight* (3)1: 15–31.

Sandiford, M. (2014), Has the death spiral for Australia's electricity market begun? *The Conversation* June 29, 2014 (accessed online 28/7/2016: https://the-conversation.com/has-the-death-spiral-for-australias-electricity-market-begun-2858).

Smith, A. and Stirling, A. (2007), Moving Outside or Inside? Objectification and Reflexivity in the Governance of Socio-Technical Systems, *Journal of Environmental Policy & Planning*, 9:3–4: 351–373.

Smith, A. and Stirling, A. (2008) Social-Ecological Resilience and Sociotechnical Transitions: Critical issues for Sustainability Governance, STEPS Working Paper 8, Brighton: STEPS Centre.

Smith, A. and Stirling, A. (2010), The Politics of Social-Ecological Resilience and Sustainable Socio-Technical Transitions, *Ecology and Society* 15(1). (accessed 23/2/16: http://www.ecologyandsociety.org/vol15/iss1/art11/).

Solar Citizens (2016a), The State of Solar: Australia's Solar Rooftop Boom, *Solar Citizens*, www.solarcitizens.org.au (accessed 28/7/16: http://roofjuice.com.au/wp-content/uploads/2016/06/FINAL-Australias-Solar-Rooftop-Boom-A4-Report150dpi.pdf).

Solar Citizens (2016b), COAG Energy Council outcomes cloudy, *Solar Citizens*, www.solarcitizens.org.au (accessed 22/8/16: http://www.solarcitizens.org.au/coag_energy_council_outcomes_cloudy).

Stewart, R. (2016), Why you should stay on the grid, even with your solar-powered batteries, *The Conversation* February 16, 2016 (accessed online 25/7/2016: https://theconversation.com/why-you-should-stay-on-the-grid-even-with-your-solar-powered-batteries-41765).

Stockholm Environment Centre (2010), Can Transition Management Contribute to Social Resilience? (video accessed 31/8/16: http://www.stockholmresilience.org/research/research-videos/2010-03-25-can-transition-management-contribute-to-social-resilience.html).

Stone-Jovicich, S. (2015), Probing the interfaces between the social sciences and social-ecological resilience: insights from integrative and hybrid perspectives in the social sciences, *Ecology and Society* 20(2): 25. (accessed 22/8/16: https://doi.org/10.5751/ES-07347-200225).

The Conversation (2016), FactCheck Q&A: is Australia the world leader in household solar power?, *The Conversation* March 28, 2016 (accessed online 19/7/2016: https://theconversation.com/factcheck-qanda-is-australia-the-world-leader-in-household-solar-power-56670).

The Economist (2017), Renewable energy – A world turned upside down: Wind and solar energy are disrupting a century-old model of providing electricity. What will replace it?, *The Economist* February 25, 2017 (accessed online 27/5/17: http://www.economist.com/news/briefing/21717365-wind-and-solar-energy-are-disrupting-century-old-model-providing-electricity-what-will).

Walker, G. and Shove, E. (2007), 'Ambivalence, Sustainability and the Governance of Socio-Technical Transitions', *Journal of Environmental Policy & Planning*, 9:3–4: 213–225.

Warren, B., Christoff, P. and Green, D. (2016), 'Australia's Sustainable Energy Transition: the Disjointed Politics of Decarbonisation,' *Environmental Innovations and Societal Transitions*, in press (accessed 31/8/16: http://www.sciencedirect.com/science/article/pii/S2210422416300016).

Wood, T. (2016), 'Australia's energy sector is in critical need of reform,' *The Conversation* July 14, 2016 (accessed online 25/7/2016: https://theconversation.com/australias-energy-sector-is-in-critical-need-of-reform-61802).

Vorrath, S. (2016), Australia's top solar states and suburbs, *Renew Economy* 26 May 2016 (accessed online 28/7/16: http://RenewEconomy.com.au/2016/australias-top-solar-states-and-suburbs-23048).

CHAPTER 4

Mapping Sociotechnical Resilience

Vivek Kant and Justyna Tasic

Introduction

This chapter presents a framework for addressing sociotechnical resilience, and uses it for analyzing the electricity sector specifically in Singapore, the most rapidly advanced city in Southeast Asia. Essentially, sociotechnical resilience is addressed by delving into the nature of sociotechnical systems as a hybrid, multiperspectival, agentic and engineered system (Amir and Kant 2018); along with, identifying the method for approaching sociotechnical resilience in terms of a proactive approach of managing resilience. Resilience has been scrutinized, contested and negotiated by several academic fields ranging from human and social sciences to engineering: safety (e.g. Hollnagel et al. 2006); social-ecological systems (e.g. Gallopin 2006; Walker et al. 2004); resilient communities (Tasic and Amir 2016); urban resilience (Wagenaar and Wilkinson 2015); coastal resilience (Klein et al. 1998); disaster

V. Kant (✉)
Industrial Design Centre, Indian Institute of Technology Bombay (IITB), Mumbai, Maharashtra, India
e-mail: vivek.kant@iitb.ac.in

J. Tasic
Institute of Catastrophe Risk Management, Interdisciplinary Graduate School, Nanyang Technological University, Singapore, Singapore

Future Resilient Systems, Singapore-ETH Centre, Singapore, Singapore

S. Amir (ed.), *The Sociotechnical Constitution of Resilience*,
https://doi.org/10.1007/978-981-10-8509-3_4

resilience (Dahlberg et al. 2015); disaster risk reduction (Pelling and Wisner 2012); and engineering resilience (Holling 1996; Park et al. 2013). While the list continues to grow, skeptics also question the "emptiness" of the term (Sehgal 2015). In contrast, we believe that by meaningfully engaging resilience in its sociotechnical manifestation will allow for prevention of system malfunctions as well as rapid recovery in the aftermath of debilitating shocks in sociotechnical system.

Resilience is often characterized as the ability of systems to recover following a shock. While this ability is present for many systems, its manifestation for sociotechnical system proves a challenge. This is due to the constitution of sociotechnical system as a special class of systems that consist of hybrid entities: intentional agents and material technologies. People appear in many roles as well as form a part of larger entities (e.g. organizations) that support the workings of the technological aspects of sociotechnical system. Thus, both people and technologies are mutually entangled—they are hybrid. For addressing these hybrid entities as systemic constructs requires characterizing them as people in their surround (the system comprehended from within) as well as parts of an engineered system at various levels of abstraction (the system comprehended from the outside). These hybrid parts are interlinked to each other and serve as interwoven pathways for information, capital and material exchange that support system functioning and prevent malfunctions. The aim is thus to provide a framework that enables the identification of tools and mechanisms for proactively addressing sociotechnical resilience from a large-scale systemic approach for identifying aspects that support resilience and removal of others that hamper resilience. Towards this end, the current chapter presents a conceptual model of a sociotechnical system and uses a mathematical network-based (more specifically, meshwork-based) analytic approach for identifying system phenomena that will aid sociotechnical resilience. Mathematical network-based analysis is not new in its application for resilience (e.g. Kapucu 2006; Arianos et al. 2009); however, in this chapter the emphasis is to use mathematical network analysis for characterizing holons and meshworks for comprehending sociotechnical resilience (Amir and Kant 2018). Further developments that present a quantified view of resilience have not been addressed in this chapter and is a topic for further discussions.

To summarize, the framework presented in the first part of the chapter will be used for a preliminary mapping of the Singapore electricity sector. This chapter is divided into two main parts. Part I presents a model for sociotechnical system and the framework for studying sociotechnical resilience. Part II uses the concepts developed in Part I for a preliminary analysis of the Singapore electricity sector. For the ease of readability, a list of abbreviations has been placed in an appendix at the end of this chapter.

Theoretical Background

Sociotechnical Systems

Sociotechnical systems present an abstruse forefront for researchers. This is due to the fact that they are hybrids: simultaneously a conglomeration of networked actors as well as a system of technical entities, designed to function correctly. This ambiguous nature has resulted in social sciences- and engineering-based views on sociotechnical systems. These streams differ in the way they conceptualize sociotechnical system (ontological view) and deploy methods to analyze sociotechnical system (epistemological view).

In providing the distinction between these two approaches, De Bruijn and Herder (2009) note that the discipline of engineering adopts a technical-rational view on systems analysis. The ontological view recognizes the presence of multiple subsystems that are interconnected. There are multiple conflicting objectives and interdependencies. The system is recognized to display dynamics at several time scales. On the basis of the ontology adopted by engineers for characterizing sociotechnical system, the epistemological view allows them to treat sociotechnical system using a rational and mathematically formal manner. Structured models, which can be tested, allow for the distinction between analysis and design as separate activities. Thus, the model presents a better understanding of the system by providing appropriate concepts to depict the interrelations between variables. In many cases, certain subsystems are modeled as black boxes in terms of input and output processes and also accounting for state transitions of these systems. A standard modeling ploy in systems theory is treating the system in terms of recursion of systems within systems. The modeler applies decomposition techniques to comprehend a class of systems, which in turn can be used for decomposing the corresponding subsystems. The analyst/modeler adopts a third-person perspective for addressing the system and its constituent subsystems. In general, the epistemology adopted is for viewing systems from the outside—a view from without.

In contrast to the engineering approach, the actor-based approach involves treating sociotechnical system in terms of actors as strategic agents involved in ongoing negotiations. These interactions are political, economic and social in nature, often comprising of multi-faceted meshworks of relations. In terms of the epistemological view, these agents are involved in power dynamics and capitalization of interests. Decision-making is an emergent based on multiple pressures that cannot be resolved by rational and mathematically based formal approaches. Sociotechnical system modeling in this case is a result of negotiation and modeling does not denote a separation of analysis and design. In many cases, decisions are viewed in terms of a sequential progression of phases that are difficult to break down in terms of individually self-contained sub-steps. In general, the epistemological viewpoint is adopted for comprehending the systems in terms of the enacted reality of the participants—a view from within. A fundamental challenge for comprehensively addressing sociotechnical resilience is to recognize the two views of approaching sociotechnical system (from without and within) as complementary and formulate the means for taking both of these together.

A second challenge is posed by the engineered dimension of sociotechnical system. It is engineered for correct functioning and avoiding malfunctions. Malfunctions of the technical aspects of sociotechnical system represents an injury, accident or catastrophe. Typically, this lies in the purview of the engineering and technically oriented personnel involved in system functioning. However, accident investigations have revealed that the social aspects of sociotechnical systems play a major role in accidents (Rasmussen and Svedung 2000). Typically, the accident is caused by a mismatch in systemic parts, often blamed on human error. However, by considering the totality of the scenario often reveals that these accidents were "waiting to unfold." This could be due to the nature of the domain of activity involving high risk (Perrow 1984) or it could be due to the market pressures that force the companies to cut labor cost or operate machinery often at boundaries of safe functioning. While multiple sources of the blame could be found, the prevention of accidents (one aspect of achieving a resilient sociotechnical system) is possible by recognizing the larger social constraints on the functioning of the technological components. Sociotechnical system involves people at multiple levels of functional abstraction ranging from operators in power plants to legislators in the governments, all of whom are vital to the sociotechnical system (Fig. 4.1). While a clear-cut causal path is difficult to be drawn from large-scale social processes to the actual functioning of the technical components, these processes can be comprehended in terms of the constraints they impose on the

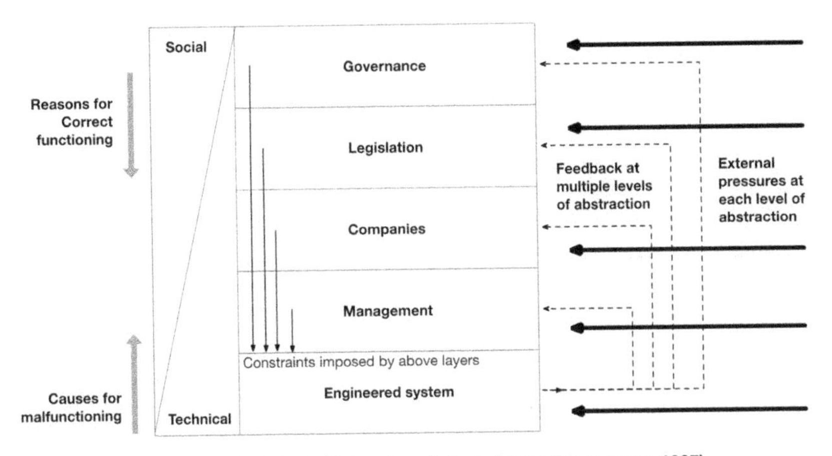

Fig. 4.1 Sociotechnical systems at multiple levels of abstraction. (Adapted from Rasmussen 1997)

actual technical activity in sociotechnical system in terms of reasons for what is deemed as correct functioning (see Polanyi 2005, Ch. 11).

Further, an additional problem is posed by the hybridity of sociotechnical system. Inclusion of social elements makes modeling of these systems different from standard engineering modeling. Typically, engineering modeling of the technical components involve careful definitions of parts that can be structurally decomposed. Functioning of such systems are based on measurable variables that could be precisely specified in terms of quantitative models and controlled experiments of the technical designs. These carefully designed aspects of systems description in quantitative terms do not acquire additional contextual functionality that emerge during operation and influence the overall development trajectory of the system. Even though it has a multitude of parts, a technical system is essentially a *complicated* system amenable to structural decomposition. In contrast, the hybrid functioning of people and technologies influences the developmental trajectory of the entire system. It is difficult for complex systems to be reduced to primal entities that are involved piece-meal. Thus, compared to *complicatedness* of engineered systems, sociotechnical systems are *complex* in nature (see Rosen 2000, p. 292 for difference between complicatedness and complexity).

The hybridity of the sociotechnical system makes it difficult to be decomposed structurally; that is, the boundary between the human and technology is blurred. The blurring of boundaries is present due to the functioning of the sociotechnical system. For example, operators use strategies to handle system functioning as well as get adapted to handling the power plant over a period of time, which may lead to malfunction. Further, engineers design technical aspects of systems, which in turn get appropriated by users for their own purposes bringing in dynamic vulnerabilities not addressed in the original design plans. Therefore, rather than structural decomposition, a functional abstraction is required to address the complexity of the sociotechnical system in order to bring the agents and immediate contexts together in a unified whole (see Rasmussen and Svedung 2000, p. 30).

To ensure resilience, analytical modeling of sociotechnical system requires discovering the functional parts (agents in their immediate contexts) in the meshwork of dynamic relations. This would ascertain the linkage of the local considerations with the global shifts in the sociotechnical system. Highlighting these dynamics will ensure that the sociotechnical system would include not only the ability to recover from a shock but also to avoid malfunctions caused by vulnerabilities imposed by the everyday functioning of the sociotechnical system.

Based on the preceding characteristics of the sociotechnical system, the first step in modeling sociotechnical system would be to begin with a qualitative analysis based on the perspective of the actors within the system. This analysis reveals hybrid wholes based on functional abstraction, as opposed to structural decomposition, for multiple levels of the sociotechnical system. The functional wholes are linked together in terms of meshworks involving information, capital and matter exchanges. From this qualitative viewpoint of the first step, the second step is formulated. In this second step, the system is viewed from the outside in terms of interactions that can be quantified using mathematical techniques. The next subsection presents the modeling of sociotechnical system based on these two steps in greater detail.

Sociotechnical Resilience: A Blended Approach

The last subsection provided the necessary basis for addressing sociotechnical resilience in terms of mathematical network-based analysis. It emphasized that sociotechnical system has to be addressed both from the within and without, in terms of functional abstraction at various levels of a multi-tiered sociotechnical system. To incorporate all these insights into a unified framework, this subsection presents two main steps for comprehending sociotechnical resilience. First, using an interpretive understanding it emphasizes the discovery of functional holons for charting purposeful behavior at the various levels of sociotechnical system. In this step, constraints imposed by the upper levels on the lower levels of the sociotechnical system are also identified. Second, this meshwork map is analyzed in terms of information, capital and matter exchange among various holons using mathematical network analysis tools.

The interpretive viewpoint is often used in the social sciences to understand the lived experience of people in their contexts from their viewpoints. The interpretive approach tries to discover how people make sense of their worlds; the meanings these people have for their everyday activity and environments; the ways in which these meanings are comprehended, contested, resisted, accepted and negotiated in a manner that ultimately shapes both the agents and their technological contexts. In this first step, data about this situated behavior is acquired for actors at various levels of the multi-tiered sociotechnical system (see Fig. 4.1, adapted from Rasmussen 1997). Actors at various levels will be different; nevertheless, they will all be involved in purposeful behavior—purposeful from their own viewpoint as well as purposeful for the overall functioning of the sociotechnical system. Therefore, in order to incorporate the qualitative data and mold it in a form amenable for network (meshwork) analysis, it will be categorized in terms of five main categories that together form a functional whole (holon) for representing purposeful behavior: act, agency, agent, scene and purpose. These five categories are derived from the literary theorist Kenneth Burke (1969). Burke used the five categories, called the Dramatistic Pentad, derived from drama, to comprehensively characterize purposeful activity. The category of act captures the activities, transactions and happenings. Scene captures where these activities were conducted, or environmental aspects, or arenas to support the act. Agent addresses the entities (persons, organizations and institutions). Agency involves the means or strategies agents attempt while involved in the activity. Finally, purpose captures the motives behind the purposeful activity. Taken together, these categories form a conceptual whole based on the interpretive understanding of the actors within a system. These unified functional wholes are identified as holons.

Holons (Koestler 1971) are wholes that are complete in themselves but at the same time are parts of a larger system. Characterizing functional holons eliminates the strict structural reduction of the systems into its subsystems as prescribed in the engineering approach. Further, it captures the actors and their meaningful context in an encapsulated form. These holons are linked to other functional wholes and the subcomponents of these holons are linked to other similar subcomponents. These interlinkages between holons serves as interconnections for information, capital and material exchanges. Thus forming an integrated and nested meshwork of holons and their subcomponents (act, agency, agent, scene and purpose). It is to be noted that while holons are presented in terms of categories that include agents, in many cases of engineered systems, agents may not be present. Therefore, in cases of technical components, the holons will comprise of the scene, purpose and agency. These three take into account the presence of technological components, the reason for their design and functioning, along with ways in which they function effectively along with information, capital and material flows.

The term meshwork, in contrast to networks, is emphasized in this framework. This is due to accounting for the processual nature of change in the sociotechnical system as well as its surrounds. Networks present a case of interaction where the lines act as connectors between already formed points/nodes. In contrast, a different and an apt need for treating the interconnections is to characterize them as meshworks (Ingold 2007, 2013). Meshworks are "interwoven trails rather than a network of intersecting routes … trails along which life is lived" (Ingold 2007, p. 75). An example in this case is the meshwork as a web of relations that support flows of information, capital and matter—paths that support the commingling of the social and the technical. These meshworks of flows connect holons as well as constantly shape the holons themselves. Thus, the identity and integrity of the holons cannot be viewed disassociated from these meshworks. More broadly, viewed in light of meshwork, holons can be characterized as knots or entanglements of these various relations. Thus, the holons and the meshwork of entanglements are mutually supported by each other and are dynamic functional wholes. Along with meshworks that mutually constitute the holons, the subcategories of holons (e.g. agent, purpose, etc.) can also be envisioned in terms of sub-meshworks and can be addressed separately in terms of the individual relations. For example, how the technological context at various levels of the sociotechnical system interacts with each other in terms of information, capital and material

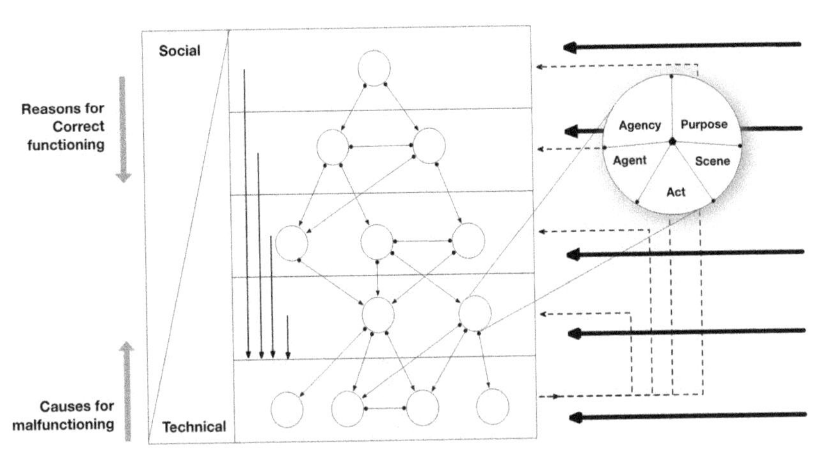

Fig. 4.2 Holons and meshworks at multiple levels of abstraction of the sociotechnical systems, constraints and feedback also presented at various levels of abstraction

flows. By addressing these dynamic relations between holons and subcategories of holons, the sociotechnical system is identified in terms of mutuality of exchanges that sustain the sociotechnical system as well as support the transformation of the system as a whole, in the aftermath of a shock. Along with the holons, the constraints imposed by each holons at the level below are also captured at each level of the sociotechnical system (see Fig. 4.2).

Once the sociotechnical system is formulated in terms of meshworks involving the holons, it becomes amenable to mathematical network analysis (MNA). Network analysis is a branch of mathematics that is based on graph theory (algebra). MNA allows for addressing interactions in a graph at the micro level to simulate behavior at the macro level. Systems design in this case would include formulating adequate holons and the networks of exchanges in order to achieve a desired macroscopic behavior of the sociotechnical system for achieving systemic resilience. At the same time, since the network of holons is dynamic in nature, constraints at multiple levels can also be changed and adjusted to understand the macroscopic behavior. Using a quantified mathematical approach allows for devising multiple simulations based on different conditions of the holons. Further, aspects of flows between the holons; identification of critical holons;

constraints on certain parts of the meshwork; and graceful degradation due to removal of holons are among some aspects that can be explored in greater detail for understanding macroscopic behavior. These simulations provide an insight into scenarios that would not have been possible to test in real life. Sociotechnical resilience could then be categorized as systems property based on changes of certain subsets of the meshworks in relation to the whole meshwork. Alternatively, sociotechnical resilience could also be characterized in terms of flows of information, capital and matter in relation to the holons. Other definitions of sociotechnical resilience based on network analysis are also possible and can be explored further. The ways in which the system recovers from shocks at a macro scale can be addressed in terms of the actual activities at the micro scale of agents and technologies, which in turn could be experimented with in order to find optimal outcomes. Thus, presenting a proactive approach for achieving resilience.

The next section demonstrates a part of the first step of the analysis outlined above. It presents the sociotechnical system of the electricity sector in Singapore and highlights the actors at each level of the sociotechnical system. This actor map will be used as the initial basis for the interpretive approach outlined in step one of the proposed analysis. Essentially a preliminary survey of holons, constraints, meshwork flows and the environmental stressors will be provided.

Electricity Sector in Singapore: A Preliminary Mapping

Brief Overview of the Electricity System in Singapore

Electric power generation and delivery is extremely crucial to support advanced economies like Singapore. In the highly urbanized and technologically advanced environment, electricity is a key factor in sustaining most of economic activities. All critical infrastructure systems are heavily dependent on electrical power, including the electricity system itself. Both the interdependencies inside of the electricity system and the environment impact on adaptation conditions that need to be fulfilled in order to maintain the stability and reliability of the electricity industry. Furthermore, the feedback loop consequences of changes affect not only the power system but also other dependent systems, such as transportation, telecommunication, health care, gas transmission and so on. Therefore, Singapore must pursue to create a conducive environment for a resilient electricity sector. The solutions have already been sought in competitive, innovative and sustainable development (clean technology and smart energy economy).[1]

Since 1995, the energy market in Singapore changed fundamentally through a series of liberalization processes.[2] The liberalization brought much more variability and competitiveness to the market. In 2001, after the implementation of a new legal and regulatory framework was introduced, the Energy Market Authority (EMA) was established to regulate both the electricity and the gas industry. The wholesale electricity market, known as the National Electricity Market of Singapore (NEMS), was gradually put in practice together with new rules and business processes imposed by Energy Market Company (EMC). In 2003, the NEMS officially was opened to competition and the wholesale electricity trading started.

Over the last decade, Singapore's electricity generation industry moved away from fuel-oil steam turbine plants by building new or adapting existing plants to new cleaner and more efficient technology—combined cycle gas turbines. Those changes influenced on the structure of imported energy resources and the rise of natural gas share in electricity generation. In 2014, total energy production in Singapore produced by main power producers accounted for approximately 46000 GWh (EMA 2015). Due to very limited indigenous energy resources, more than 95% of Singapore's electricity was generated using imported natural gas (EMA 2015). Most of the gas was piped to Singapore from Indonesia (two pipelines) and Malaysia (two pipelines). Currently, there are six main electricity generation companies that produce more than 90% of electricity in Singapore (EMA 2015). Three of the biggest generators, namely Senoko Energy Supply (SES), Tuas Power Generation, and YTL PowerSeraya, generate more than 60% of electricity (EMA 2015). The main electricity retailer in Singapore is SP Services (33% of the market share), which is as well the only retailer to households and small consumers (below 2000 kWh of contestability threshold). Furthermore, there are six other main electricity retailers selling to contestable consumers. Among them, the largest was Senoko Energy Supply with almost 15% of the market share (EMA 2015). In the last decade, the overall electricity consumption grew from 35,000 to 46,000 GWh (EMA 2015). Industry, commerce and service, and households consume most of the electricity in Singapore.

This preliminary analysis of the Singapore electricity system was prepared on the basis of secondary data sources available online for the public. The data were gathered through in-depth exploration of the various websites of ministries, agencies and organizations, the addresses of which are included in the Online Sources section at the end of this chapter. Among the analyzed data were: national policy framework documents; strategic committee reports; Energy Market Authority (EMA) regulatory documents (market rules), statistical data, and informative documents; ministries statements and information; organizational structures, corporate governance information, annual company reports; and press releases.

Singapore Electricity System as a Sociotechnical System

Singapore's electricity system is conceptualized as a multi-tiered sociotechnical system that consists of six abstraction levels: government, regulatory agency, organization (company), management, staff and engineered systems (Fig. 4.3). The levels change in their predominance gradually from the emphasis on the social fabric at the top to technical materiality at the bottom. On the one hand, from the top down, the government and the regulatory agencies impose constraints on the lower levels providing the reasons for its correct functioning. On the other hand, causes of malfunction proceed bottom up. Functional holons (constituting actors) and their interconnecting meshwork of information, capital and matter exchanges are identified at each level abstraction. The actor's significance describes the actor's role in securing resilience of the electricity sector operations, both in normal operation conditions as well as during emergencies. The meshworks' flow intensity demonstrates the load of the exchanged information, capital and matter. As such, the information and capital interlinkages predominate in the upper and middle levels, and material interconnections prevail in the bottom tiers of the model.

Government

The first tier of the power system model (Fig. 4.3) comprises mostly ministries involved in constitution of laws, governance of the country, and shaping the national policy related to the electricity sector. The most central is the Ministry of Trade and Industry (MTI), which frames National Energy Policy and leads the Energy Policy Group (EPG) (MTI 2007). Furthermore, the Ministry of the Environment and Water Resources (MEWR) and the Ministry of Foreign Affairs (MFA) are also very important actors, as they are involved in EPG coordination (MTI 2007). In addition to the ministries, the National Research Foundation (NRF), as a

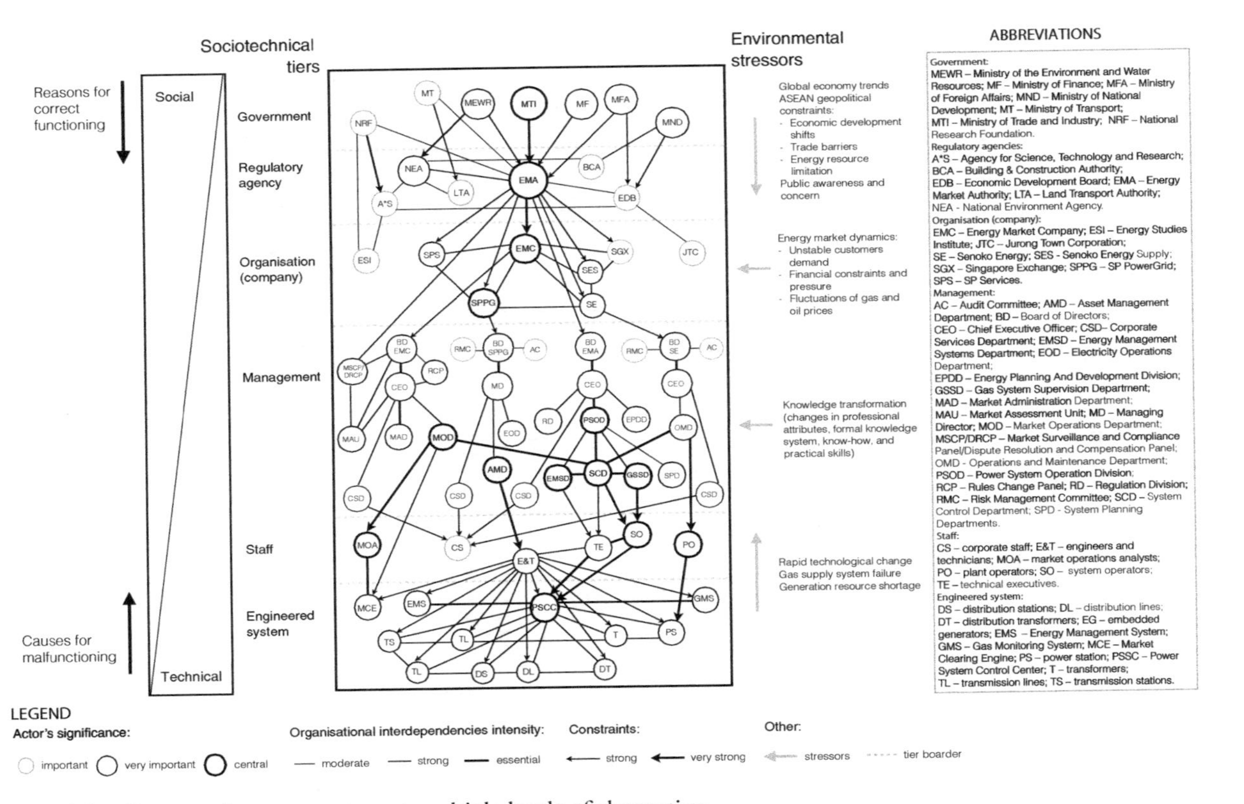

Fig. 4.3 Singapore's power system at multiple levels of abstraction

part of the Prime Minister's Office, is an important actor that establishes national research and development strategy.[3]

Regulatory Agency

The regulatory agency level embraces the statutory boards of the Singapore government and the research agency. The boards perform an operational function by announcing regulations, standards and plans. At this layer, the most influential actor is the Energy Market Authority (EMA), as it is the main energy sector regulator, operator and developer (EMA 2010). Agency for Science, Technology and Research (A*STAR) is a leading actor for research and development in the electricity sector.[4]

Organization

The third tier represents organizations and companies that are involved in the most important operations of the National Electricity Market of Singapore (NEMS): the market company, generation companies, retailers, the stock exchange, the transmission company and the market support services licensee.[5] Additionally, Energy Studies Institute (ESI)—responsible for energy research and development programs[6]—and Jurong Town Corporation (JTC)—developer and manager of industrial facilities[7]—were included. The interwoven meshwork pathways of generation companies and retailers are explicated with the example of Senoko Energy (SE), the biggest electricity generator in Singapore (EMA 2015), and the subsidiary retail company—Senoko Energy Supply (SES). At this level, the most central actor is the Energy Market Company (EMC), which as a market operator governs and secures all of the market operations, including the electricity dispatch instructions, calculation of the Uniform Singapore Energy Price (USEP) for consumers, nodal prices for generators and the safety reserve. The meshwork flows are very strongly integrated between EMC, SE and the transmission company (i.e. SP PowerGrid (SPPG)), and they comprise mostly information and capital exchanges. The capital flows, stimulated by EMC, are crucial for organizing the wholesale operations.

Management

The management level demonstrates the actors involved in decision-making and execution, operation reviews, establishing development strategies, risk management, and contingency and emergency preparedness plans. For transparency reasons, only the management actors of EMC, SPPG, EMA and SE are presented, as we found them indispensable for the system's security and reliability. At this tier, the meshwork paths consist of information and capital relations. The most central actor is the power system operator, Power System Operation Division (PSOD) in the EMA, which monitors and controls electricity generation and transmission, and gas transmission (EMA 2010).[8] The interconnections of the PSOD's departments are fundamental for performance of Singapore's whole power system, both in normal operation conditions and during emergency (EMA 2016). In particular, the most essential is the information on market forecast and system load forecast, and dispatch instructions are exchanged between the System Control Department (SCD) and the Gas System Supervision Department (GSSD), the Market Operations Department (MOD) in the EMC, and the Operations and Maintenance Department (OMD) in the SE.[9] Furthermore, the SCD is connected through a key information interlinkage with the Energy Management Systems Department (EMSD) that is accountable for the operation of the mission-critical computer systems. Another central actor is Asset Management Department (AMD) in the SPPG, which is responsible for the management and maintenance of the electricity transmission and distribution network (Singapore Power 2010).

The emergency planning and response of the power system is managed by the PSOD, which develops and maintains the necessary system operating procedures to manage a power system crisis (EMA 2010). The System Planning Departments (SPD) prepare and maintain the Singapore Electricity Emergency Plan and the Singapore Power System Restoration Plan. Additionally, during the states of high risk and emergency the PSOD acquires additional autonomy to control the power system (EMA 2016). The high-risk state is a pre-emergency state, when the PSOD can temporarily increase the level of control to bring the system back to the normal state. When the system enters the emergency state, returning to the normal state requires non-dispatchable load[10] and allows high PSOD control on the power system.

Staff

The staff layer represents the employees and teams that are the most involved in the power system operations and maintenance. At this layer, a single circle represents a whole professional group, for example, plant operators. The staff actors are limited to those related with the crucial actors on the management tiers. At this layer, the most central actors are system operators (SO), plant operators (PO) and market operations analysts (MOA) due to their higher responsibility for the power system, power plant and electricity market operations. The SO are essential actors as they regulate the power system by electricity dispatch schedules and supervise the gas transmission system. The system operators work 24/7, with the support of technical executives, to ensure electricity system functioning.[11] The PO control the power stations operation. Moreover, the engineers and technicians (E&T) are also influential actors, as they are responsible for design, analysis and maintenance of particular parts of the technical system.

Engineered System

The engineered system level consists of technical systems that facilitate electricity generation, transmission and distribution. These systems constitute the subsystems of a large technical system—the entire power system. At this layer, the meshwork exchanges involve mainly the matter and information flows. The most central actor is the Power System Control Centre (PSCC), the nerve center of Singapore's power system, located in Ayer Rajah substation. The PSCC is supported by mission-critical computer systems, that is, Gas Monitoring System (GMS), Energy Management System (EMS).[12] From the center, the SO is able to control and monitor in real time all electricity generation plants and the dispersed transmission network in Singapore.

Holons and Meshworks in the Singapore Electricity System

Earlier in part 1 of this chapter, holons were described as encapsulated entities that take into account various actor's relations in its immediate contexts. Thus, the power system was characterized as a meshwork of holons and its resilience is a result of the holons and meshwork interdependencies. Specifically, holons were characterized as consisting of subcategories at each level of abstraction: (1) agent, (2) act, (3) agency, (4) scene and (5) purpose. For instance, the MTI's holon (Fig. 4.4) comprises the following: (1) agent: MTI; (2) act: coordination of Energy Policy Group (EPG)[13] and regulation of national energy policy; (3) agency: National Energy Policy Framework of Singapore; (4) scene: political and economic

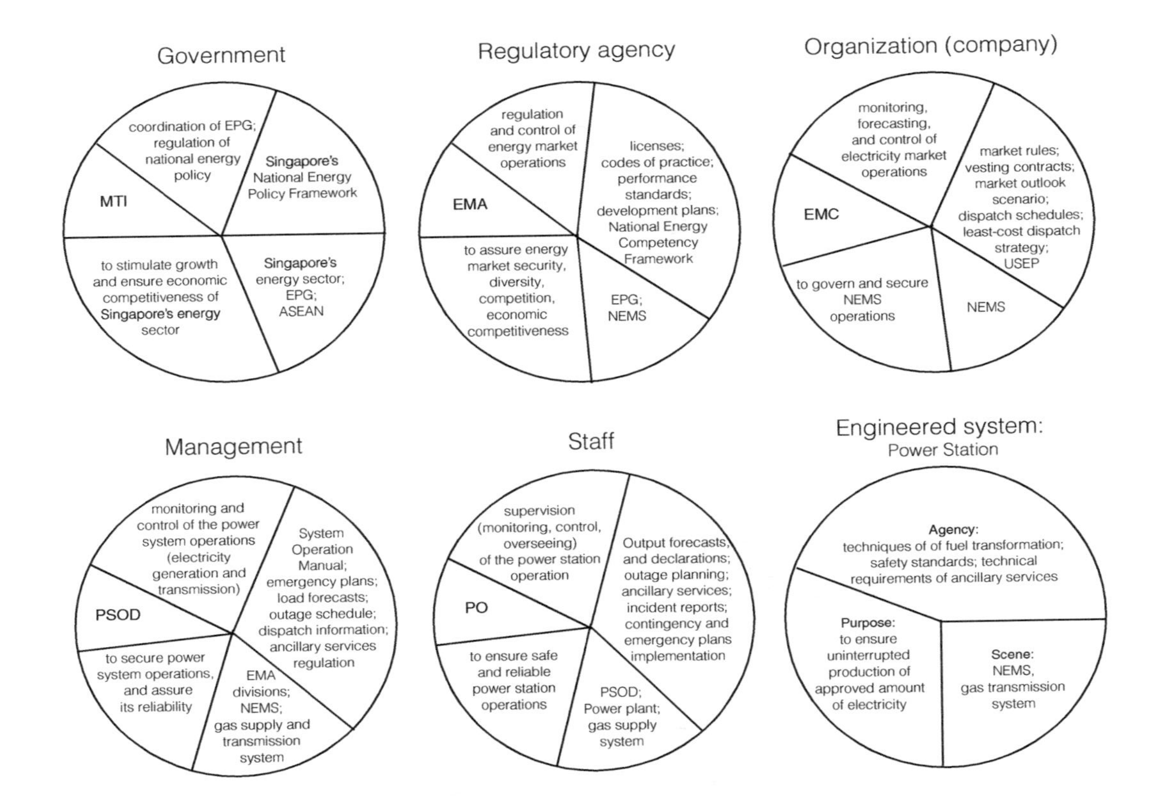

Fig. 4.4 Examples of holon's subcategories at various levels of abstraction

relations in Singapore's energy sector, EPG, and ASEAN member countries; and (5) purpose: to stimulate growth and ensure economic competitiveness of Singapore's energy sector.

Constraints

Moreover, our preliminary analysis revealed many inter-tier constraints imposed by the upper levels on the lower levels. Starting from the top-most level, there are strong interaction links between the ministries and regulatory agencies. Both the ministries and the agencies have a tight-knit cooperation regarding establishing the national policy framework and strategies that provide the reasons for correct functioning of the electricity system as a whole. As such, the agencies, through the statutory mission, implement the national framework and constitute a mediator between the organizational and governmental layers. Therefore, there are also strong constraints imposed by the regulatory agencies upon the organizations. Particularly, the EMA, as the main energy market regulator, develops the energy policies and plans, issues the operation licenses, establishes codes of practice, performance standards, and thus, has a strong influence on the organizations involved in the energy sector (EMA 2010). Next, in terms of company policy and strategy implementation, the organizations impose constraints on the related management actors, usually firstly through the board of directors (BD), advised by respected committees, and CEO. The management actors, such as departments and divisions, in the execution work, provide instructions to the staff. For example, risk and performance management of the power system in Singapore is firstly organized by the Energy Planning and Development Division (EPDD)[14] that develops and aligns the energy policies and plans to ensure that security, economic competitiveness and environmental sustainability of the power system are maintained.[15] Further, the SPPG through the BD SPPG, supported by the Risk Management Committee (RMC), incorporates the energy policies and plans into the corporate and risk-management strategies. Next, the AMD, in terms of multiple aspects, retrains the E&T, who preserve the power system's technical subsystems within the given frames (practicality, safety and cost limitations). Furthermore, there are very strong, and important in terms of the system security, constraints imposed by the SO and the TE over the nerve center of system operations (PSSC).

Environmental Stressors

In addition, as presented in Fig. 4.3, the power system in Singapore is under the pressure of dynamic environmental stressors that propagate vertically through the various layers or horizontally through a particular layer. For example, the stressors at the first two levels (government and regulatory agency) propagate top down and impact the whole system. Among the distinguished stressors are global economy trends, Association of Southeast Asian Nations (ASEAN) geopolitical constraints (economic development shifts, trade barriers and energy resource limitation) and public awareness and concern.

Singapore's economy is strengthened by active involvement in worldwide economic and political cooperation. However, the regional relations in Southeast Asia are particularly important for Singapore. More specifically, the collaboration among the ASEAN is fundamental for development of Singapore (Ng 2012).[16] That is, the disturbances in the member countries and conflicts between them could hinder the economic and political conditions in Singapore. As such, the regional relations influence on Singapore's electricity sector security, which is highly dependent on the imported resources. Particularly, it is important to realize the high interdependency between electricity generation and natural gas supply. In 2014, more than 95% of electricity in Singapore was produced from natural gas (EMA 2015), thus, as the past has shown, the interruptions in the gas supply cause serious disruptions in the energy production (Chang 2015). However, through building up the LNG terminals,[17] Singapore wishes to become an important and more independent actor in the Trans-ASEAN electricity network.[18] Currently, approximately 22% of total natural gas imports came in the form of LNG (EMA 2015). In addition, Singapore, being a part of the United Nations, also declared a carbon efficient pathway for economic growth and a 36% reduction in the emissions intensity by 2030.[19,20] The Paris Agreement is assumed to bring changes in the electricity sector, which is expected almost entirely to switch to natural gas for power generation and increase Smart Grid investments.[21,22]

Furthermore, civil society in Singapore is an active partner in the governance processes; thus, the people's opinion and concerns are taken into account in solution-oriented decision-making and planning. To bridge a gap between the government and the society, Singapore established Community Development Councils that aid local administration of governmental policies and implement community programs. At the organization level, the pressures are mostly related to the electricity market dynamics. Especially after the energy market liberalization steps taken in 2003, the organizations in the electricity sector operate in the aggressive and competitive market environment where the unstable customers demand, financial constraints and pressure bring the burden of the cost-effective priorities.[23] Furthermore, the organizational operations and decision-making are highly dependent on the prices of gas and oil; thus, any fluctuations will have a direct impact on their actions. The management and staff tiers are under pressure of the knowledge transformation stressor. Technological progress and dynamic market conditions bring emerging constant need for professional competency metamorphoses. Thus, as such, the knowledge transformation conveys redefining professional attributes, formal knowledge, know-how and practical skills. Finally, the engineered system is constrained by rapid technological change, generation resource shortage and gas supply system failure. Causes of malfunctions propagate bottom up beginning from lowermost levels in the electricity system. In this case, starting from the engineered systems and affecting the upper layers.

Conclusion

This chapter aimed to provide a two-step framework for proactively addressing sociotechnical resilience and its applicability for Singapore's electricity sector. The proposed analytic approach responded to the challenge of different social and engineering views of approaching sociotechnical system (from within and from without) and provided the complementary means to address both of them. The novel blended approach of addressing sociotechnical system enables the identification of structures that support resilience and removal of others that hamper resilience mechanisms. As such, the sociotechnical system is introduced as a multi-tiered abstract construct where the hybrid sociotechnical entanglements and interwoven flows of information, capital and material exchanges are deeply rooted at each level of the abstraction. Towards this end, from a large-scale systemic approach, the conceptual model of sociotechnical system is portrayed as a meshwork of holons.

Based on an exploratory case study of Singapore's electricity sector, the chapter introduced the first step of the proposed sociotechnical resilience framework. In this matter, it presented a multi-tiered sociotechnical system model of the power system and highlighted the actors at each level of abstraction. The preliminary analysis of the various actors allowed the possibility of discovering encapsulated holons, tracing meshwork flows and identifying constraints in the system. Furthermore, the investigation of environmental stressors propagating throughout the system sheds light on the external conditions that affect the system, and thus, influence on its ability to recover from a shock and avoid malfunctions. This chapter introduced preliminary actor and meshwork identification for comprehending sociotechnical resilience. A detailed situated qualitative analysis and mathematical network analysis has not been yet developed. These aspects will be the next step of our project on comprehending the sociotechnical resilience of Singapore's electricity sector.

Acknowledgement This work is an outcome of the Future Resilient Systems project at the Singapore-ETH Centre (SEC), which is funded by the National Research Foundation of Singapore (NRF) under its Campus for Research Excellence and Technological Enterprise (CREATE) program (FI 370074011).

Notes

1. Report of the Economic Strategies Committee, sections: Ensuring Energy Resilience and Sustainable Growth, Growing Knowledge Capital, available at: www.mof.gov.sg/Portals/0/MOF%20For/Businesses/ESC%20Recommendations/ESC%20Full%20Report.pdf (Accessed: 15.03.2016).
2. About the NEMS, available at: www.emcsg.com/AboutTheMarket(Accessed: 15.03.2016).
3. NRF, Clean Energy Programme, available at: https://rita.nrf.gov.sg/ewi/cepo/default.aspx (Accessed: 15.05.2016).
4. A*STAR, Cleantech Projects (e.g. Intelligent Energy Distribution Systems; Experimental Power Grid Centre) available at: www.a-star.edu.sg/Industry/Industry-Sectors/Cleantech.aspx (Accessed: 15.05.2016).
5. EMC, Market Players, available at: www.emcsg.com/marketplayers (Accessed: 15.05.2016).
6. ESI, Mission and Values, available at: http://esi.nus.edu.sg/about-us/mission-and-values (Accessed: 15.05.2016).
7. For example, Jurong Rock Caverns, available at: www.jtc.gov.sg/industrial-land-and-space/pages/jurong-rock-caverns.aspx (Accessed: 15.05.2016).
8. Power System Operation Division, available at: /www.ema.gov.sg/Power_System_Operation_Division.aspx (Accessed: 15.05.2016).
9. Executive management team at Senoko Energy, available at: www.senokoenergy.com/executive-management-team (Accessed: 15.05.2016).
10. In the normal state, the PSO shall issue dispatch instructions as the primary means of coordinating the real-time operation of the PSO controlled system.

11. EMA, Power System Operator, available at: www.ema.gov.sg/Power_System_Operator.aspx (Accessed: 15.05.2016).
12. EMA, Mission-critical systems, available at: www.ema.gov.sg/Essential_Systems.aspx (Accessed: 15.05.2016).
13. The EPG formulates and coordinates Singapore's energy policy and strategies.
14. More specifically, the Policy and Planning Departments.
15. EMA, Energy Planning and Development Division, available at: www.ema.gov.sg/Energy_Planning_and_Development_Division.aspx (Accessed: 15.05.2016).
16. The relations gained even higher meaning after establishment of the ASEAN Economic Community in December 2015, which assumes economical integration, cohesion and cooperation (e.g. free flow of goods and services, investments, capital and skills).
17. Singapore LNG Corporation, available at: www.slng.com.sg/website/index.aspx (Accessed: 15.03.2016).
18. The network's plans comprise the Trans-ASEAN Gas Pipeline and the ASEAN Power Grid.
19. The reduction percentage is compared with 2005 levels.
20. MFA, Statement about The Paris Agreement, available at: www.mfa.gov.sg/content/mfa/overseasmission/geneva/press_statements_speeches/2015/201507/press_20150703.html (Accessed: 15.03.2016).
21. EMA, Smart Grid investment plans, available at: www.ema.gov.sg/cmsmedia/Newsletter/2012/04/eyeon-emaIES.html (Accessed: 15.03.2016).
22. EDB, Clean Energy, available at: www.edb.gov.sg/content/edb/en/industries/industries/clean-energy.html (Accessed: 15.03.2016).
23. Strait Times, Article about the fforecastedfull electricity market liberalization, available at: www.straitstimes.com/singapore/electricity-market-to-be-fully-liberalised-in-2018 (Accessed: 15.03.2016).

References

Amir, S., & Kant, V. (2018). Sociotechnical Resilience: A Preliminary Concept. Risk Analysis, 38: 8–16. https://doi.org/10.1111/risa.12816.

Arianos, S., Bompard, E., Carbone, A., & Xue, F. (2009). Power grid vulnerability: A complex network approach. Chaos: An Interdisciplinary Journal of Nonlinear Science, 19(1), 013119.

Burke, K. (1969). A grammar of motives. Berkeley: University of California Press. Original published in 1945.

Dahlberg, R., Johannessen-Henry, C. T., Raju, E., & Tulsiani, S. (2015). Resilience in disaster research: three versions. Civil Engineering and Environmental Systems, 32(1–2), 44–54. https://doi.org/10.1080/10286608.2015.1025064

De Bruijn, H., & Herder, P. M. (2009). System and actor perspectives on sociotechnical systems. IEEE Transactions on Systems, Man, and Cybernetics-Part A: Systems and Humans, 39(5), 981–992.

Chang, Y. (2015). Energy in Singapore: The engine of economic growth from 1965 to 2065. In Quah, E. ed. (2016). Singapore 2065: leading insights on economy and environment from 50 Singapore icons and beyond. (pp. 343–347). New Jersey: World Scientific.

Energy Market Authority (2015). Singapore Energy Statistics 2015. Retrieved from https://www.ema.gov.sg/cmsmedia/Publications_and_Statistics/Publications/SES2015_Final_website_2mb.pdf (Accessed 05.05.2016).

EMA (2010). Introduction to the National Electricity Market of Singapore. Retrieved from https://www.ema.gov.sg/cmsmedia/Handbook/NEMS_111010.pdf (Accessed 23.04.2016).

EMA (2016) Chapter 5. System Operation, in Singapore Electricity Market Rules. Retrieved from https://www.emcsg.com/f283,7842/Chapter_5_System_Operation_1Apr16.pdf (Accessed 05.05.2016).

Gallopín, G. C. (2006). Linkages between vulnerability, resilience, and adaptive capacity. Global Environmental Change, 16(3), 293–303.

Kapucu, N. (2006). Interagency communication networks during emergencies boundary spanners in multiagency coordination. The American Review of Public Administration, 36(2), 207–225.

Klein, R. J. T., Smit, M. J., Goosen, H., & Hulsbergen, C. H. (1998). Resilience and Vulnerability: Coastal Dynamics or Dutch Dikes? The Geographical Journal, 164(3), 259–268.

Koestler, A. (1971). The ghost in the machine. (2nd.print.). London: Pan Books. Original published in 1967.

Holling, C. S. (1996). Engineering Resilience versus Ecological Resilience. In P. C. Schulze (Ed.), A common perceptual parameter for stair climbing for children, young and old adults (pp. 31–44). Washington, D.C.: National Academies Press. Retrieved from http://www.nap.edu/read/4919/chapter/4

Hollnagel, E., Woods, D. D., & Leveson, N. (2006). Resilience engineering concepts and precepts. Aldershot, England; Burlington, VT: Ashgate.

Ingold, T. (2007). Lines: a brief history. London; New York: Routledge.

Ingold, T. (2013). Making : anthropology, archaeology, art and architecture. London; New York: Routledge.

MTI (Ministry of Trade and Industry) (2007). Singapore's National Energy Policy Report – Energy for Growth. Retrieved from https://www.mti.gov.sg/ResearchRoom/Documents/app.mti.gov.sg/data/pages/885/doc/NEPR%202007.pdf (Accessed 23.03.2016).

Ng, W. H. (2012). Singapore, the energy economy : from the first refinery to the end of cheap oil, 1960–2010. London; New York: Routledge.

Park, J., Seager, T. P., Rao, P. S. C., Convertino, M., & Linkov, I. (2013). Integrating Risk and Resilience Approaches to Catastrophe Management in Engineering Systems. Risk Analysis, 33(3), 356–367. https://doi.org/10.1111/j.1539-6924.2012.01885.x

Pelling, M., & Wisner, B. (2012). Disaster risk reduction: Cases from urban Africa. Routledge.

Perrow, C. (1984). Normal accidents : living with high-risk technologies. New York: Basic Books.

Polanyi, M. (2005). Personal Knowledge: Towards a Post-Critical Philosophy. London: Routledge. Original published in 1958.

Sehgal, P. (2015). The Profound Emptiness of "Resilience." Retrieved from http://www.nytimes.com/2015/12/06/magazine/the-profound-emptiness-of-resilience.html?_r=0

Rasmussen, J. (1997). Risk management in a dynamic society: a modelling problem. Safety Science, 27(2), 183–213.

Rasmussen, J., & Svedung, I. (2000). Proactive risk management in a dynamic society. Karlstad, Sweden: Swedish Rescue Services Agency. Retrieved from https://www.msb.se/ribdata/filer/pdf/16252.pdf

Rosen, R. (2000). Essays on life itself. New York: Columbia University Press.

Singapore Power (2010). Annual report 2010. Enhancing Capabilities Powering Growth. Retrieved from http://www.singaporepower.com.sg/irj/go/km/docs/documents/SP%20Content/Sites/Singapore%20Power/Site%20Contents/Annual%20Reports/documents/2010/SP_AR2010.pdf (Accessed 23.03.2016).

Tasic, J., & Amir S. (2016). Informational Capital and Disaster Resilience: The Case of Jalin Merapi, Journal of Disaster Prevention and Management, 25(3), 1–18.

Wagenaar, H., & Wilkinson, C. (2015). Enacting Resilience: A Performative Account of Governing for Urban Resilience. Urban Studies, 52(7), 1265–1284. https://doi.org/10.1177/0042098013505655

Walker, B., Holling, C. S., Carpenter, S. R., & Kinzig, A. (2004). Resilience, adaptability and transformability in social--ecological systems. Ecology and Society, 9(2), 5.

PART II

Disrupted Environments

CHAPTER 5

Weather-Ready Nation or Ready-Weather Agency? Emphatic Accuracy and Sociotechnical Resilience in the National Weather Service

Jen Henderson

Introduction

In a presentation at the American Meteorological Society's annual meeting in January 2016, U.S. National Weather Service (NWS) administrators updated an audience of forecasters and researchers on efforts to "evolve" their agency. Rather than focusing the majority of their efforts on increasing forecast accuracy through technoscientific means, they commissioned an Operations and Workforce Analysis from McKinsey & Company to evaluate their current state of affairs, identify service gaps, and offer a new philosophy to the community. The ensuing report suggested that the NWS should prioritize "deep relationships" with specific groups, called "core partners," who could "amplify their [hazard] messages" and offer "focused [forecast] products" tailored to partner needs.

J. Henderson (✉)
University of Colorado, Boulder, CO, USA
e-mail: Jennifer.henderson-1@colorado.edu

S. Amir (ed.), *The Sociotechnical Constitution of Resilience*,
https://doi.org/10.1007/978-981-10-8509-3_5

Just how forecasters ought to proceed and which activities should be deployed by local operational offices to this end, however, remained unclear (Swanson-Kagan et al. 2016). To date, these strategies are being worked out. Still, what became apparent during that presentation is that the agency's strategic plan, "Weather-Ready Nation Roadmap," and its core initiative, Impact-Based Decision Support Services (IDSS), would be key to their efforts. Based on "actionable ideas," IDSS would promote resilience, defined as a flexible "set of skills" that would allow the agency to "continually evolve over time to meet the changing needs of society" (ibid.). In its efforts to build an adaptable organization, responsive to disruptions from disasters, they are not alone.

Communities, organizations, and government agencies have increasingly mobilized the concept of resilience in disaster discourse to develop strategies that might help people to return to a state of normalcy after catastrophes, or adapt to ongoing fluctuations in their environments (International Strategy for Disaster Reduction 2009). Yet, as this edited collection demonstrates, resilience is a polysemous concept that eludes straightforward definitions in its deployment. Successful analyses of resilience, as the introduction argues, require a hybrid approach to understanding the infrastructures, technologies, humans, and practices—the sociotechnical nature—of our systems. One important aspect of resilience that is addressed less often in disaster discourse is its normative aims, or as Healy and Mesman note, "the resilience of what, for whom, and at what cost or tradeoff…?" (2014, 155–56). Such questions help us understand the motives and consequences of efforts that might otherwise be deemed innocuous or remain hidden. Such questions may also help reveal moments of possible intervention. In this chapter, I use this normative frame to extend Healy and Mesman's dimensions of resilience to a sociotechnical context as it emerges in the NWS Weather-Ready Nation Roadmap, hereafter NWS Roadmap.

Amir and Kant (2018) argue that resilience is not fully deployed in these types of sociotechnical systems without hybridity, an inclusion of the social and the technical acting together. Many formulations of resilience, they write, "lack conceptual theorization of what constitutes resilience for sociotechnical systems" because "definitions and conceptualizations seem to be taken for granted" (2017, 2). The NWS warning and forecast system is certainly sociotechnical in its mutual entanglement of humans and technologies, especially at the operational level where multiple networks of people and machines together produce information about the weather. It does so on a daily basis, though during times of hazardous weather, additional networks and systems—emergency management and state or federal agencies, for example—are enrolled. A sociotechnical resilience, then, is more reflective of the efforts the NWS deploys to "realize a Weather-Ready Nation" (National Oceanic and Atmospheric Administration 2013b, 1).

To more fully articulate the mutual entanglement of technical and non-technical, I examine how sociotechnical resilience performs in the NWS Roadmap, specifically through a collection of activities and concepts, practices and policies instantiated in one core initiative, IDSS. Together they function as a sociotechnical system that shapes forecasters' engagement with their various publics through the technological elements of the weather prediction process and creates a nation of citizens who have become mobilized as "weather ready." It is a future based on what Amir and Kant call a collection of anticipatory practices (2018), which propel a sociotechnical system toward a particular vision of the future. In this case, it is one in which the NWS itself is adaptable to different threats and its citizenry prepared.

A reading of sociotechnical resilience through a normative lens offers insight into its multiple valences, which I show inflect two categories of concern. One valence of sociotechnical resilience points to the external threat of extreme weather events that destroy lives and livelihoods and requires costly recovery. In this instance, sociotechnical resilience resembles a more common deployment of the term in that the NWS will leverage IDSS to build an America that is "aware of" and "responsive" to dangerous weather. Another valence reflects threats internal to the NWS, which arise when those in private industry and politics call into question the value of the government-funded public forecaster. IDSS, then, can be seen as a strategy to bolster the agency's relevance and value in society. Each valence of sociotechnical resilience promotes different images of the forecaster and who the forecaster ought to be in relationship to their professional identity, their scientific enterprise, and their ethical commitments to different publics. In this chapter I offer the term *empathetic accuracy* as a way to reframe what appears as a binary of accuracy and relationships in forecaster work to better reflect the complex practices and relational negotiations they encounter in their daily practices. I define the term as a hybrid ethic that imbricates care, sociotechnical resilience, and accuracy in the practices, attitudes, and materials technoscientific experts develop for their publics (Henderson 2016). Viewed through empathetic accuracy, IDSS, then, has the potential to dramatically reshape the future role of the NWS forecaster in society and to ensure their agency's efforts are more equitable and just.[1]

Critical Participation and Normative Dimensions of Sociotechnical Resilience

Asking normative questions about sociotechnical resilience and its implications for local communities requires an analyst to reveal alternative ethics that might reshape a sociotechnical imaginary (Jasanoff 2015b) of the forecaster, especially those that better align with their agency's imperative to protect lives. Like other scholars in Science and Technology Studies (STS) who reveal the normative dimensions of technologies (Heidegger 1977; Doppelt 2000; Jonas 2009; Jasanoff 2015a), I suggest the ethics of intervening sits alongside the analyst's assessment of those scientific or technological endeavors examined. In this effort, I perform what Downey and Zuiderent-Jerak call "meta-activism" (2016) as I offer the NWS suggestions for how I believe they might move forward in the twenty-first century. In particular, I posit that the NWS "scale up and make visible" (Downey 1998; Downey and Dumit 2009) an image of the forecaster that reflects matters of care and relationality already present in and important to sociotechnical resilience in forecaster work.

Why examine an initiative like IDSS and why now? Scholars in STS have long argued that nascent sociotechnical systems flex and shift as relevant social groups negotiate their activities and practices (Ravetz 1971; Hughes 1987; Bijker et al. 1987; Latour 2007). IDSS is no exception. Not yet a standardized practice, it is malleable and thus open to critique and redesign—a clear reason to engage with it at this time. However, as Winner notes, it is not just the processes leading to closure that should occupy the analyst's efforts. Instead, our work can, and perhaps should, pay attention to the consequences of such practices and the invisible groups who have had no say in its production (Winner 1993). In the case of IDSS, I suggest the agency's focus on a limited group of "core partners" disappears the lay public, leaving these public safety officials as a proxy for multitude of publics.

This chapter is based on my participant observations as a facilitator and subject matter expert for eight IDSS webinars. Administrators at the NWS created this online platform as an internal forum for their meteorologists to explore and negotiate definitions and boundaries of IDSS. In what follows, I give an overview of the NWS and its concept of resilience, broadly conceived; I then offer an overview of the NWS Roadmap and three classifications of IDSS activities that differently deploy sociotechnical resilience on the ground; next I highlight the external and internal inflections of this concept in the NWS Roadmap; finally, I conclude by arguing for an *empathetic accuracy* as a new reading of the sociotechnical resilience reflected in IDSS, one that better aligns forecasters with their mission to protect lives.

The National Weather Service and Accuracy

Whenever hazardous weather is likely to occur somewhere in the United States, operational meteorologists located in one of 122 local NWS forecast offices across the country assess predictive information; create a host of "products" that give spatial, temporal, and explanatory details about what might happen (e.g., a daily forecast for rain); and provide "services" (often in the form of a product) to a variety of publics. Paid for through taxes and managed by the government agency, the National Oceanic and Atmospheric Administration, forecasters issue products at no cost and, often, in a standardized form. More tailored forecast information can be purchased from private-sector meteorologists who own proprietary software and technologies that facilitate specific client needs (National Research Council 2003). NWS publics are divided into two categories: public safety officials called "core partners" or "customers," and the lay public, whom they alternately refer to as "users," "the public," or "society." During the minutes during which storms form and weather hazards emerge, NWS forecasters create products called warnings, which are alerts that not only classify and visually demarcate dangerous weather but also give general advice on actions people should take. Since 1870, NWS activities have been a part of the commitment that reflects the agency's primary responsibility "to protect lives and property" (National Oceanic and Atmospheric Administration 2013a).

National Weather Service products also represent an administrative and public demonstration of forecasting skill, which the agency measures based on the correctness of the prediction's timing, location, threat type, and severity. In this context, accuracy "refers to the general or unspecified predictive value of a forecast or forecasting method" to an individual or community (National Research Council 1999, 12). Accuracy became a primary driver of an agency-wide effort to update forecasting technologies in the 1980s and 1990s, an effort called the Modernization and Associated Restructuring Demonstration, or Modernization. Its goal was "to improve the accuracy, timeliness, and efficiency of its weather forecasts and warnings..." and, to that end, the agency spent, "... over $4 billion to modernize NWS weather observing, information processing, and communication systems" ("Weather Forecasting: Systems Architecture Needed for National Weather Service Modernization" 1994). Discourses of forecast accuracy during the latter twentieth century evoked technoscientific improvements, as evidenced by the focus of its funding and outcomes of its policies. Around the same time, others posited that accuracy is important to forecasting only if someone is able to successfully use the prediction (Murphy 1993). Later, the 2005 NWS strategic plan, for example, highlights users of weather information as central to measures of accuracy: "As we focus on improving our services and expanding their scope we will consult effectively with all who are affected by our services and be guided by our customer's needs" (National Weather Service 2005, 6).

In many ways, accuracy continues to function as a dominant principle in the agency, motivating its sociotechnical resilience. Radar and satellite networks, real-time observational instruments, forecaster workstations, computer databases, and computer models—together these instantiate promises of accuracy to "improve" or "advance" warnings and they often drive congressional funding, policy changes, and forecaster practices (The National Academies 2012; *The Weather Service Modernization Act* 1992). An emphasis on public service and decision support exists in agency reports as well, though it is less visible and more ambiguous. In fact, some have pointed out that the NWS tends to substitute accuracy for the goal of serving people (Morss 2005). Thus, developing more accurate technologies and robust infrastructures often stands in for relationships with their many publics, at least until recently.

As with many kinds of disasters, those that involve extreme weather become moments of possible transformation for institutions. In 2011, more than 600 people died in two tornado disasters in the United States. The first struck six states in the Southeastern United States on April 27, producing 363 tornadoes and causing 340 deaths (National Weather Service 2011a). Dubbed the Super Outbreak, it was followed the next month by another series of storms in the Midwest, which produced an EF-5 tornado in Joplin, Missouri, on May 22. This deadly tornado ripped through the town on a Sunday afternoon, destroyed 25 percent of the structures, and killed 159 people, making it the deadliest single tornado since 1948 (*NWS Vision for IDSS* 2016). Death tolls like this had not occurred since the 1970s, just before the current warning infrastructures had been put in place. Questions posed by forecasters in the wake of these disasters centered on a problematic of public response: How could so many die when the science of meteorology and its technologies were so accurate and timely?

In light of such questions, the NWS launched its Weather-Ready Nation strategic plan in late 2011, and a more detailed Roadmap in 2013, to help execute a new agency vision (Furgione 2012) to "become more agile, open and flexible [in order] to achieve its mission to meet society's changing needs" (2013b, 1). The Roadmap is a seventy-five-page document with four subdivided areas that describe activities, technologies, and milestones the NWS must meet to successfully implement its vision by the year 2020, at which time they "will [have] translated the Strategic Plan into real-life actions that save lives and livelihoods" (U.S. Department of Commerce 2011). Importantly, the agency acknowledges some limitations of accuracy, noting that they "must go beyond the production of accurate forecasts and timely warnings and build in improved understanding and anticipation of the likely human and economic impacts of such events" (2013b, 5). Rather than solely pursue advances in accuracy, the agency would deploy its forecasters as instruments of sociotechnical resilience to discern through unspecified activities and processes the complexities of their publics. As will be discussed next, the strategic plan and the initiative it creates represent the agency's efforts to generate particular normative visions for what counts as threats to the system and how those affected might "bounce back" from disruption.

Resilience: Generally and Locally

The relationship between resilience and vulnerability is particularly complex. Traveling out of systems ecology in the 1970s, through various disciplines—human geography, psychology, hazards literature, and disaster studies—and into engineering and community planning, resilience arrives at each destination ambiguous and multiple (Miller et al. 2010; Endress 2015). Current framings typically cast resilience as a positive notion, one that proposes institutions, communities, buildings, and ecosystems should be altered to resist or adapt to stressors and to rebound to normal. Within a system, those who face the greatest losses, expressed as an inability to recover or most susceptible to harm, are considered vulnerable and in need of strategies of resilience. Vulnerability, then, is a "relational notion" (Healy and Mesman 2014, 155) to resilience, one likewise arising in the 1970s in disaster contexts to highlight those least capable of performing flexibility in the system and who might leave it open to shocks and stressors.

Together, vulnerability and resilience constitute an anticipatory philosophy, one of worst-case scenarios, though just which threats one should prepare for and when the work of resilience should complete is unclear (Endress 2015). Authors Healy and Mesman (2014) suggest that the normative implications of this framework get "overlooked" amid the "imprecision" of its ambiguity. In the high-stakes contexts of disaster work, failures to grapple with the ways vulnerability and resilience are constructed and the consequences for people on the ground may result in higher body counts. In the daily work of weather forecasting where population vulnerabilities often translate into fatalities, this is especially true, which gives ethical weight to a plan designed to protect lives.

Protecting lives is a two-pronged endeavor. A primary function of the NWS Roadmap, the authors write, is to articulate a strategy that builds "increasing community resilience for future extreme events" (2013b, 5). On the one hand, flexibility and adaptability should occur daily within forecasting work as meteorologists diagnose and issue predictions for the weather. The plan notes that the NWS will

> evolve from a paradigm where the forecaster generates products based on static definitions [of threats] toward a services model where the forecaster works closely with core partners to recognize their needs and provide expertise to community decision-makers. (2013b, 11–15)

In this case, sociotechnical resilience transforms forecasters' labor from a surveillance of atmospheric conditions and agency metrics to an administration of relationships with others. This form of care constructs a new knowledge of risk that de-centers their scientific expertise in favor of shared epistemic authority.

On the other hand, sociotechnical resilience should propagate through the larger warning and forecasting system, which is comprised of other agencies and infrastructures, where it will be taken up by communities affected by weather hazards. The NWS Roadmap expects lay publics to shift awareness of and activities toward identified dangers as part of a regime of preparedness. Yet, "not all communities" the report admits "are adequately equipped to prepare for, protect from, or cope with environmental hazards and their impacts" (2013b, 41). In this regime, those who are unaware of such dangers or unable to act "appropriately" to prevent harm to themselves are sought out, studied, and incorporated into frameworks of the social sciences for identifying and mitigating possible injustices (Adger 2006; Cutter 2006). To address such vulnerabilities, however, may mean that forecasters prescribe decisions that such individuals ought to make in circumstances forecasters have determined to be life threatening. In this regime, IDSS is the vehicle through which much of this vision will be deployed. Like many sociotechnical imaginaries, it shapes for different communities "how life ought, and ought not, be lived" (Jasanoff 2015a).

Broadly defined, IDSS is a sociotechnical system of practices and people, things and bodies that materially affect prediction strategies. For the NWS, it is the "overarching paradigm" of the agency that holds in tension multiple notions of resilience as shared fears and shared futures. A central fulcrum in this tension, IDSS is defined as a "provision of relevant information and interpretative services to enable core partners' decisions when weather, water, or climate has a direct impact on the protection of lives and livelihoods" (2013b, 65–68). Much of the Roadmap is a working out of the institutional vocabulary and apparatuses that might lead to its successful implementation. However, metrics of success or failure in the NWS Roadmap are largely absent, as are specific activities that forecasters will use to perform IDSS with public safety partners and other bureaucratic agencies.

As the name Impact-Based Support Services implies, forecasters will continue to focus on developing accurate products and on understanding the multiple ways that local weather affects people and structures, called "impacts." The notion of "support services," then, suggests that forecasters will offer guidance to core partners and learn their unique challenges. Such support reformulates forecasting practices around people rather than mere predictive skill. A forecaster, the report notes, must understand "what and how weather impacts a decision from the core partner's perspective," which then allows them to communicate "uncertainty in understandable terms" (2013b, 12). More importantly, then, "decision support" situates judgments about people's lives as a crucial site of expertise. Such negotiations require flexible definitions of risk and distributions of authority in evaluating what constitutes an impact or a threat.

Becoming sociotechnically resilient as an institution requires forecasters to reframe their expectations that accuracy should be a primary driver of prediction. IDSS situates meaningful rapport with partners as central to sociotechnical resilience. Thus, forecasters construct better informational relations, which reinforce questions about "the content of the information and form how it is meant and how it will be used" (Amir and Kant 2018, 12). In this way, the NWS Roadmap reveals the complex negotiations over forms of harm and disruption, which get embedded in material and social practices of IDSS.

Weather-Ready Nation Roadmap: IDSS as Sociotechnical Resilience

The NWS Roadmap articulates elements of sociotechnical resilience in each of its four subsections: Services, Workforce Evolution, Science and Technology, and Business. Each is intended to facilitate IDSS, which is framed through these subsections as a linchpin of success in a larger effort to "evolve the culture of the NWS" (2013b, 12). Evolution, in fact, is an important metaphor of sociotechnical resilience—the word "evolve" occurs twenty-seven times throughout the document, indicating a systemic flexibility that allows incremental change. And this transformation occurs on many fronts, from implementation of new tools and technologies, workforce and staffing structures, to communication of information. Invoking a kind of Darwinian notion of survival characterized by adaptation to new environments and competition through selective processes, the metaphor of "evolve" also implicates society as one that is well "prepared for and responds to weather-dependent events" (2013b, 4).

The Services subsection formulates the broad skills forecasters will perform that are oriented toward user decision making, including the interpretation, communication, and improved usefulness of weather information: "Local offices will evolve from product generators to expert decision support resources," and "to do even more to help America make better decisions" (2013b, 12, 6). As the "overarching paradigm" from which the NWS will deliver its services, IDSS becomes an obligatory passage point (Callon 1986) in this strategic plan, permitting forecasters to anticipate users' informational needs, shaped according to unique risk "thresholds" (2013b, 6). Yet, NWS efforts oftentimes deploy a public deficit model (Bucchi 2008) of relationships aimed at refiguring problematic decisions about risk, such that people—from partners to publics—make "better" decisions. That is, while IDSS represents an NWS endeavor to listen more than speak, its deployment teeters on matters of authority and control.

In the Workforce subsection, the Roadmap emphasizes ways the new role of the forecaster will align with an "evolved" NWS, which it positions differently from the staffing needs of Modernization. Rather than attend to the scientific elements of their work that might increase the accuracy of predictions—the main objective of Modernization—staff will adapt as requirements change. Central to their skillset is an ability to translate challenging elements of the forecast, such as uncertainty, for partners and to "incorporate social science principles in day-to-day operations," especially insights from communication (2013b, 34–42). Better communication, the Roadmap suggests, facilitates deeper relationships with and knowledge of partner processes; for their publics, outcomes are less complex, limited mainly to creating opportunities for improved messaging about threats. The forecaster functions as an attuned instrument to interpret people on the ground rather than the weather in the sky. Yet, lists of skills within the NWS Roadmap mainly emphasize a host of technological training, including GIS, computer modeling, data visualization, and "broader environmental science skills." Likewise, training involves physical sciences, "emerging science and technology," management and leadership, and outreach (2013b, 35–37). Little is said about the processes forecasters must learn to negotiate these "deep" relationships. Importantly, such an array of technoscientific skills seem to reflect those of the agency concerned about its own significance in society. The report notes, "It is vital that NWS remain agile to keep up with the changing science and technology and remain relevant to its evolving core partner and user requirements" (2013b, 35). Evolving the workforce ensures continued survival of the agency. Together the Services and Workforce subsections take more than two-thirds of the document.

The last two subsections, Science and Technology and Business, are the briefest in the Roadmap but highlight the scientific authority that forecasters bring to the prediction process; they likewise lay out justifications for the economic value of the NWS. Future technoscience developments are focused around innovations in infrastructural elements of the predictive process, especially computer models, observing systems, and research to operations to research (R2O2R) test bed mechanisms. They emphasize efforts to "synchroniz[e] societal impacts and environmental data" by developing "precise and accurate environmental knowledge" that can be "delivered on demand" and in forms that are "relevant to core partners' preparation, response and recovery actions" (2013b, 48–51). It is these efforts, the reader learns, that will enable forecasters to develop a "comprehensive understand of the societal vulnerabilities" and thus will ensure the "forecast value chain" continues to endure (ibid.). Here, science and technology imbricate with economics and human survival to legitimate the agency's importance to society.

It is the Business subsection that moves toward this goal most explicitly. The NWS Roadmap's objective to evolve, the document notes, is "deemed most important to the future health of the organization," a goal premised on the value of sustainability through agility (2013b, 65). Here IDSS is articulated most clearly in the Roadmap, positioned as a mechanism that can facilitate "good health" (2013b, 65–68) for the NWS. That is, IDSS will make more economically valuable forecasters' information since it will directly help others make decisions that "save lives and protect property." Business and technoscientific epistemologies produce a more robust set of "anticipatory practices" that construct "regular activities aimed at anticipating possibilities of what would occur in the future" (Amir and Kant 2018, 13): in this case, a robust agency and prepared society. Just which elements of society benefit is a point I'll examine shortly.

What exactly is IDSS as envisioned by forecasters? Just as the official definition is broad and vague, its deployment on the ground is slippery, expanding and contracting as local forecast offices illustrate their situated understandings of the initiative. From my participation in a yearlong series of webinars hosted internally by the NWS for its own employees, I have noted a diverse breadth and scope of practices that forecasters label IDSS; however, most fall into one of three categories. Although I have categorized activities to demonstrate that certain types are extensions of an older emphasis on accuracy, for forecasters in the IDSS webinars these activities are fluid in their practices but central in their focus: scientific authority. "IDSS is rooted in quality forecasts based on sound science," one forecaster noted during his presentation.

The first classification includes what I call *dissemination IDSS*, or those that require the forecaster to repackage or explain weather information for particular users. The aim is to communicate expert knowledge of weather conditions and impacts and is often unidirectional by design. Common examples include weekly webinars that offer descriptions of upcoming changes in weather and impacts through PowerPoint slides. In the days leading up to dangerous weather, these are often coupled with conference calls held with subgroups of public safety officials, such as principals of school districts. The main IDSS requirement is an attention to information design and an allowance for partner questions at the end.

Another category is *embedded IDSS*, or those that require a forecaster join a core partner group on-site to offer updated predictive information relevant to the others' needs. Perhaps the oldest and most familiar example is the iMet, or incident meteorologist, who has trained in fire weather. These NWS staff members become part of the local incident command team and are tasked with keeping firefighters safe "by interpreting weather information, assessing its effect on the fire and communicating it to fire crews" (National Weather Service n.d.). While iMets are traditionally dedicated only to fire contexts, NWS forecasters are increasingly following the iMet model in other hazardous weather situations, inserting themselves in emergency operation centers and with other government entities, like those preparing for significant community celebrations like Mardi Gras, for example. Accuracy of forecast information still sits at the center of these activities, though the dimensions of accuracy—location, timing, and expressions of uncertainty—are tailored to the context.

The last category I call *relationship IDSS*, which differs from the previous ones in its emphasis not on the meteorological expertise forecasters deploy but on the concerns of the people with whom they work. Relationship building exists in the other two categories, but is not the explicit goal—it is an outcome or byproduct of the situation and interactions. Relationship IDSS, then, is not about distributing predictive information just before or during hazardous weather; instead, it is performed continuously throughout the year (e.g. through scenario planning and partner workshops) and it is done in ways that allow for an exchange of perspectives and a deeper understanding between individuals about their respective roles and requirements in the warning process. In this type of IDSS, trust built between experts and publics shapes the practices forecasters engage in and the types of predictive information they produce. It is this latter category that has the most potential for highlighting the multiple ethics of accuracy and care that exist in and are most important to future forecaster practices.

External Sociotechnical Resilience: Ready, Responsive, and Resilient

I would like to return to Healy and Mesman and their call for those of us who employ the term resilience to ask questions about the context and consequences of its use. Within the scope of the NWS Roadmap and the activities and practices generated by the IDSS initiative, I suggest that sociotechnical resilience is inflected primarily in two ways: internally toward the agency and its staff and externally toward core partners. Distinctions between the two are often muddled in execution since similar practices embody both internal and external articulations against a threat.

The NWS Roadmap projects an imaginary of the United States population that is "ready, responsive, and resilient"—an external inflection of sociotechnical resilience.

> Weather-Ready Nation is about building community resilience in the face of increasing vulnerability to extreme weather and water events. In the end, emergency managers, first responders, government officials, businesses, and the public will be empowered to make faster, smarter decisions to save lives and protect livelihoods. (National Oceanic and Atmospheric Administration 2011b, 1)

To achieve national success, the Roadmap mobilizes its argument at the smaller scale of local communities, with specific emphasis on partners and decision makers. Motivation for this framing is not only an increase in extreme weather occurrences but a growing vulnerability each year within the population and infrastructures to those extremes. External resilience might thus be conceived of as resilience for individuals, especially vulnerable individuals, and is deployed against threats of dangerous weather.

Just who or what is vulnerable and the mechanisms by which vulnerability ought to be identified is not clearly addressed in the NWS Roadmap. This is perhaps not surprising. Vulnerability is an empty signifier in a risk society (Beck 1992) where experts are not always able to identify who is likely to be most negatively affected by disasters. Expanding the network of expertise to public safety officials via IDSS and through social science research, NWS administrators suggest, enables forecasters to better survey where vulnerabilities exist in the public and how choices might be made to mitigate any weaknesses. Yet, the specific sociotechnical resiliences that ought to be generated and where remains nebulous.

External sociotechnical resilience is likewise framed as a support mechanism for decision makers who are responsible for taking actions on behalf of the lay public. By building meaningful relationships with core partners, NWS forecasters can shape a resilience against their own ignorance of the core partners' circumstances, contexts, needs, or thresholds. Similarly, these same core partners—emergency managers and other governmental agencies, school districts, and hospital administrators—manage resources that are important to public safety. One assumption is that forecasters and partners together generate sociotechnical resilience against similar threats of weather hazards and loss of life and so bear joint responsibility for outcomes. As one report notes, IDSS is an "enhanced, multi-disciplinary approach that will empower emergency managers, first responders, government officials, businesses and the public to make fast, smart decisions to save lives and livelihoods" (American Meteorological Society 2012, 2). IDSS, then, directs responsibility for outcomes to core partners who make decisions about community needs and so seem to bear sole responsibility for consequences of their efforts.

This assumption is supported by my own observations. NWS forecasters frequently lead core partners in assessing the weather threat but stop short of weighing in on the actual course of action. As happens many times, forecasters' support of a decision ends with their best accounting of the threat's timing, location, duration, and magnitude, and attending uncertainties. How partners elect to act on the predictive information that forecasters provide, the NWS implies, falls outside the scope of their role as scientific experts, an important tension given the trust implicit in relationship building. For example, a common response from forecasters to public school administrators' who request advice on whether or not to close schools early or release buses filled with students is, "That's not my job." In one way, then, external resilience is also deployed against the threat of direct responsibility for a choice of action and its outcomes. Still, forecasters can be implicated in decisions that affect the larger community through possible ambiguities in their explanations of uncertainty, as well as obstacles to clear expressions of confidence. IDSS, then, potentially offers the NWS insulation from accountability related to their predictive efforts.

Internal Sociotechnical Resilience: Weather-Ready Nation or Ready-Weather Agency?

Internal sociotechnical resilience might be also conceptualized as one that strengthens the agency against the threats from others. Of particular note are dangers of irrelevancy raised by Congress as budgets become more constrained (Rosenfeld 2000; Samenow 2015). Vulnerability, then, is not just a characteristic of lay publics but of organizations. To this end, the NWS Roadmap notes that forecasters must "add value to objectively produced information" from computer models, making them indispensable to partners and thus to the national economy (2013b, 56). Sociotechnical resilience, as articulated in the praxis of forecasting through IDSS, is about making the institution resistant to external financial pressures and valuable to society and, thus, competitive for federal funding. Science and accuracy are still important, but internal sociotechnical resilience, or self-preservation-cum-relationships, appears to be on equal footing.

One challenge to this internal inflection is the response by individual forecasters in the NWS. Many fear that IDSS represents a threat to their scientific identity as they transition more toward "interpreters" and communicators of predictive information. As one representative forecaster said to me during my field work:

> How much hand-holding do we need to do when we tell people how to understand the forecast? I think we may be going a little too much into that. And the IDSS stuff is lots and lots of hand-holding. Are we doing science? Call it something else.

As computer models have increasingly outperformed them in the accuracy of their daily predictions, forecasters have begun to ask what will become of their profession if they no longer manage their science. They are accustomed to functioning as lone authorities in matters related to weather prediction. As Phaidra Daipha notes, how such "disciplined improvisation is to be transformed into a masterful weather forecast falls under forecasters' sole responsibility and discretion" (2015, 794). It is within the metrics of this role that forecasters are most comfortable and where they take the most pride in their work.

To shift the role of the forecaster from prediction to communication is, as some have said to me, to "waste forecasters' scientific training," situating them as "consultants" for their partners and "translators" rather than experts. To these forecasters, IDSS is a strategy of sociotechnical resilience for the bureaucratic institution only; to them, it is a short-sighted sacrifice of scientific knowledge and expertise. One novice forecaster expressed her concern this way:

> I mean you talk about automating the extended [forecast], automating the TAFS [for airports], automating the GRIDS [that underpin the website forecasts], automating the short-term [forecasts]. I'm like well, if you automate everything, you need fewer people sitting here... It's like, oh, hopefully we don't automate ourselves out a job.

One tradeoff for the NWS in their Roadmap, then, is a potential devaluing and deskilling of a profession. This reading of internal sociotechnical resilience points to IDSS as a threat created by their own institution against which NWS forecasters must struggle for their scientific expertise, and in the end, the existence of their profession.

In the weather forecasting community, many suggest that regardless of implications to the role of the forecaster, the kinds of sociotechnical resilience promised by IDSS practices is worth these costs. It will result in reduced deaths of people from dangerous weather since forecasters will be better positioned to represent their expertise to the people who depend on it. But will it? Is building relationships with partners enough to ensure that NWS administrators will not have to ask again, as they did in 2011, why so many people died in a disaster in spite of current improvements in their products and infrastructures?

An implicit assumption of internal and external resilience unexamined in the Weather-Ready Nation Roadmap is that the benefits of IDSS—an understanding of partner needs—will trickle down to offer protection to the lay public. This assumption plays out in many IDSS webinars created by forecasters, which highlight the desire for partners to "amplify" their message through "broader dissemination of warnings" to different publics. However, I argue this vision of IDSS disappears the lay public in favor of a paradigm that leaves responsibility for the safety of individual citizens to the collective efforts of core partners and citizens themselves. There is no "public IDSS" for the lay public, one that extends to them a commitment to develop relationships and knowledges about the multiplicity and complexity of individual lives. I argue there ought to be. In its current form, IDSS determines criteria for hazards and mechanisms of life-saving alerts. Thus, different publics may yet face new dynamics of vulnerability from the consequences of NWS sociotechnical resilience deployed.

Several new warning infrastructures and technological tools are being designed to mimic the IDSS concept to some degree, allowing partners to give input in defining hazards. As some designers have suggested, these efforts are an improvement over older models where the forecaster's knowledge, decisions, and skill dictated the kind of information disseminated. One computer scientist explained possible future predictive structures as a matter of user preference:

> It's a concept of pull versus push once you have this national hazard database. The paradigm we're in right now is definitely a push, where the forecaster generates everything and pushes it out. But you can then imagine that a public could come to this national database and say, "Well, I'd like to see things in this way and that way," and pull things from it in different formats or different configurations.

Just how the lay public's needs or desires would become legible to such a system is unclear. There is no facilitation of comprehensive and diverse public input into the system, nor are there mechanisms for forecaster practices to better reflect individual lay public needs. Instead of embedding values derived from relationships with their communities, then, these systems potentially re-inscribe certain types of expertise and only a partial fulfillment of forecasters' ongoing mission to protect all members of their publics.

Conclusion: The Emergence of an Empathetic Accuracy

My critical participation as a social scientist in this weather community has compelled me to take seriously a normative obligation to identify problematics, images, and identities important to NWS forecasters (Downey and Dumit 2009; Downey and Zuiderent-Jerak 2016). For in shaping how they see themselves and others, I am joining them in their commitment to help protect lives by evaluating a system that ought to reflect the needs and concerns of people in harm's way. Like others who have followed this normative turn in STS, (Cohen and Galusky 2010), I believe my analysis ought to reveal to those with whom I participate—and to those who join me in disaster STS activist research—alternatives that reframe, scale up, (Downey and Dumit 2009) and reformulate future alternative roles for forecasters in society. That is, I view my activism as one of "studying up" (Nadar 1972; Priyadharshini 2003) in order to help the NWS as an organization become more attuned to and successful at meeting its ethical obligation to protect lives. As a government agency, the NWS ought to recreate a profession of scientifically trained public servants not only to be caretakers of accuracy but to be caretakers of all people equally. These two lines of care are not mutually exclusive.

Here I return to the term "empathetic accuracy," which I believe provides the NWS with an ethic that focuses on predictive precision *through* forecasters' commitment to a particular relationality with their publics. Based on principles underpinning Virginia Held's ethic of care (Held 2006), empathetic accuracy calls attention to the ways care and accuracy are imbricated in forecasters' science, co-constituted in their technological developments and policies, and reflected in the activities the direct their interactions with different publics. Mine is an ethic that highlights relationships among people and the work of meeting the needs of others as central to existence. In this sense, this hybrid ethic is not far off from the direction the NWS Roadmap guides the agency. Yet empathetic accuracy goes further. It makes visible an obligation to all publics, revealing in a new dominant image (Downey 1998) of the forecaster their relationships as a central tenet of forecasting science itself and forecasting practices in particular. I suggest it is a possible image of the sociotechnical resilience they seek (Henderson 2016).

What would a world of forecasting look like should the NWS scale up empathetic accuracy? In the short term, it could include immediate and systemic transformations in the education and training of forecasters. From the very beginning of their careers, they would have skills that match the expectations of a job based both on meteorological knowledge and public service. A diversity of coursework would include communication, for example, but not just one discipline. Future forecasters would take classes in the sociocultural and historical contingencies of meteorology, spend time grappling with the philosophical and ethical implications of their profession, and learn more about the vulnerabilities that people face through volunteer work and internships. They might model their curricula after those in other professions, like medicine, that synthesize accuracy and care. In short, their education and training should help them be more engaged in knowledge construction that is in line with new expectations for knowing how best to build relationships with others.

In the long term, deploying an empathetic accuracy could remake forecasters' relationships with a variety of publics. Rather than functioning as a largely publically invisible enterprise, the NWS might engage their publics more directly and visibly. Activities could include holding public debates, enjoining different populations to offer input on technological developments, and inviting diverse communities to help them discover what science has remained "undone" (Frickel et al. 2010). Vulnerabilities might then be recognized and addressed more quickly in their practices and missing publics might be identified and given voice.

This alternative world might also lead to a greater sense of shared responsibility for public safety, where forecasters and core partners are conjoined in their decisions and consequences. A Weather-Ready Nation might not be fragmented by a political ecology of prediction (Donner 2007) that creates inequalities even as it generates a sense of inclusion; instead, to be "weather ready" would entail a unified effort, more in keeping with those narratives of climate change that challenge clear distinction between public and expert in a world in which everyone, as with Beck's risk society (Beck 1992), is affected.

Finally, a new kind of science might emerge from this empathetic accuracy, one that takes more seriously the commitment to understanding the complexity of individual needs. It would be a science in keeping with the strong objectivity of feminist scholars who argue that better science comes from below, from a multiplicity of standpoints, and by beginning with the people on the margins (Sarah Harding 1991; Longino 2004). As Sandra Harding argues, "The scientific/epistemological and ethical/political are inseparable in standpoint approaches to research" (2009, 193). A meteorological science could be remade from the outside in, with multiple perspectives blurring lines between expert and public. It is through these kinds of revisions, I suggest, that the forecasters'—and our—ethical commitments to society can best be fulfilled.

In this chapter, I have focused much of my critique on revealing multiple, problematic readings of sociotechnical resilience as it is currently deployed in the NWS Weather-Ready Nation Roadmap and its IDSS paradigm. And I have argued that scaffolding future efforts on empathetic accuracy, a hybrid ethic that reflects important values currently circulating within forecaster practice, offers the a more promising way forward for the NWS to achieve their mission to protect lives. A sociotechnical resilience embedded with deep relationships with multiple publics, from public safety partners to everyday citizens, retains potential as a productive concept for a weather community. IDSS does some of this work but does not yet go far enough. Recognizing the problematic ways that current conceptualizations of sociotechnical resilience functions in the discourses of the NWS is a first step toward revising the system.

Acknowledgment This work was funded, in part, by a fellowship from the Graduate Student Visitor Program at the National Center for Atmospheric Research.

Note

1. There is an important distinction in the forecasting community between public forecasters employed by the U.S. government to collect meteorological information and issue free products, and private-sector forecasters who work in industry and charge a fee for their services. My work focuses only on the former.

References

Adger, W.N. 2006. "Vulnerability." *Global Environmental Change* 16 (3): 268–81.

American Meteorological Society. 2012. "State of the Weather and Climate Enterprise." Society Report. Boston, MA.

Amir, Sulfikar, and Vivek Kant. 2018. "Sociotechnical Resilience: A Preliminary Concept." *Risk Analysis* 38 (1): 8–16.

Beck, Ulrich. 1992. *Risk Society: Towards a New Modernity*. London: Sage Publications.

Bijker, Wiebe E., Thomas P. Hughes, and Trevor J. Pinch. 1987. *The Social Construction of Technological Systems: New Directions in the Sociology and History of Technology*. Cambridge, MA: MIT Press.

Bucchi, Massimiano. 2008. "Of Deficits, Deviations and Dialogues: Theories of Public Communication of Science." In *Handbook of Public Communication of Science and Technology*, edited by Massimiano Bucchi and Brian Trench, 57–76. New York: Routledge.

Callon, Michel. 1986. "Elements of a Sociology of Translation: Domestication of the Scallops and the Fishermen of St Brieuc Bay." In *Power, Action and Belief: A New Sociology of Knowledge?*, 196–233. London: Routledge.

Cohen, Benjamin R., and Wyatt Galusky. 2010. "Guest Editorial." *Science as Culture* 19 (1): 1–14.

Cutter, Susan. 2006. *Hazards Vulnerability and Environmental Justice*. Earthscan Publications Ltd.

Daipha, Phaedra. 2015. "From Bricolage to Collage: The Making of Decisions at a Weather Forecast Office." *Sociological Forum* 30 (3): 787–808.

Donner, W. R. 2007. "The Political Ecology of Disaster: An Analysis of Factors Influencing U.S. Tornado Fatalities and Injuries, 1998–2000." *Demography* 44 (3): 669–685.

Doppelt, Gerald. 2000. "What Sort of Ethics Does Technology Require?" In *Symposium on Questioning Technology by Andrew Feenberg*, 5: 157–95. Albuquerque, NM: Journal of Ethics.

Downey, Gary Lee. 1998. *The Machine in Me: An Anthropologist Sits Among Computer Engineers*. New York: Routledge.

Downey, Gary Lee, and Joseph Dumit. 2009. "What Is Engineering Studies for?: Dominant Practices and Scalable Scholarship." *Engineering Studies: Journal of the International Network for Engineering Studies* 1 (1): 55–76.

Downey, Gary Lee, and Teun Zuiderent-Jerak. 2016. "Making and Doing: Engagement and Reflexive Learning in STS." In *STS Handbook*, edited by Ulrike Felt, Rayvon Fouche, Clark Miller, and Laurel Smith-Doerr, 3rd ed., 223–52. Cambridge, MA: MIT Press.

Endress, Martin. 2015. "The Social Constructedness of Resilience." *Social Sciences* 4: 533–45.

Frickel, Scott, Sahra Gibbon, Jeff Howard, Joanna Kempner, Gwen Ottinger, and David J. Hess. 2010. "Undone Science: Charting Social Movement and Civil Society Challenges to Research Agenda Setting." *Science, Technology, & Human Values* 35 (4): 444–73.

Furgione, Laura. 2012. Weather Ready Nation and Social Sciences.

Harding, Sandra. 2009. "Standpoint Theories: Productively Controversial." *Hypatia* 24 (4): 192–200.

Harding, Sarah. 1991. "What Is Feminist Epistemology?" In *Thinking from Women's Lives.*, 105–38. New York: Cornell University Press. http://www.iep.utm.edu/fem-epis/.

Healy, Stephen, and Jessica Mesman. 2014. "Resilience: Contingency, Complexity, and Practice." In *Vulnerability in Technological Cultures: New Directions in Research and Governance*, edited by Anique Hommels, Jessica Mesman, and Wiebe E. Bijker, 155–77. Cambridge, MA: MIT Press.

Heidegger, Martin. 1977. "The Question Concerning Technology." In *The Question Concerning Technology and Other Essays*, 3–35. New York: Harper & Row.

Held, Virginia. 2006. *The Ethics of Care: Personal, Political, and Global.* Oxford, England: Oxford University Press.

Henderson, Jennifer. 2016. "'To Err on the Side of Caution:' Ethical Dimensions of the National Weather Service Warning Process." Dissertation, Blacksburg: Virginia Polytechnic Institute and State University.

Hughes, Thomas P. 1987. "The Evolution of Large Technological Systems." In *The Social Construction of Technological Systems*, edited by Wiebe E. Bijker and Thomas P. Hughes, 51–82. Cambridge, MA: MIT Press.

International Strategy for Disaster Reduction. 2009. "UNISDR: Terminology on Disaster Risk Reduction." Geneva, Switzerland: United Nations.

Jasanoff, Sheila. 2015a. "Future Imperfect: Science, Technology, and the Imaginations of Modernity." In *Dreamscapes of Modernity: Sociotechnical Imaginaries and the Fabrication of Power*, edited by Kim Sang-Hyun and Sheila Jasanoff, 1–32. Chicago: University of Chicago Press.

———. 2015b. "Imagined and Invented Worlds." In *Dreamscapes of Modernity: Sociotechnical Imaginaries and the Fabrication of Power*, edited by Sheila Jasanoff and Kim Sang-Hyun, 363. Chicago: University of Chicago Press.

Jonas, H. 2009. "Technology and Responsibility." In *Readings in the Philosophy of Technology*, 173–84. Lanham, MD: Rowman & Littlefield Publishers, Inc.

Latour, Bruno. 2007. *Reassembling the Social: An Introduction to Actor-Network-Theory*. Oxford: Oxford University Press.

Longino, Helen E. 2004. "How Values Can Be Good for Science." In *Science, Values, and Objectivity*, edited by Peter Machamer, 127–42. Pittsburgh: University of Pittsburgh Press.

Miller, Fiona, Henny Osbahr, Emily Boyd, Frank Thomalla, Sukaina Bharwani, Gina Ziervogel, Brian Walker, et al. 2010. "Resilience and Vulnerability: Complementary or Conflicting Concepts?" *Ecology and Society* 15 (3). :// WOS:000283867400035.

Morss, Rebecca E. 2005. "Problem Definition in Atmospheric Science Public Policy: An Example of Observing-System Design for Weather Prediction." *Bulletin of the American Meteorological Society*, February, 181–91.

Murphy, Allan H. 1993. "What Is a Good Forecast? An Essay on the Nature of Goodness in Weather Forecasting." *Weather and Forecasting* 8 (June): 281–93.

Nadar, Laura. 1972. "Up the Anthropologist: Perspectives Gained from Studying up." In *Reinventing Anthropology*, edited by Dell Hymes, 284–311. New York: Pantheon Books.

National Oceanic and Atmospheric Administration. 2011a. "NWS Central Region Service Assessment: Joplin, Missouri, Tornado—May 22, 2011." Service Assessment. Kansas City, MO: U.S. Department of Commerce.

———. 2011b. "Weather Ready Nation: NOAA's National Weather Service Strategic Plan." Strategic Plan. Department of Commerce.

———. 2013a. "History of the National Weather Service." Government site. National Weather Service. 2013. http://www.weather.gov/timeline.

———. 2013b. "Weather Ready Nation Roadmap." Government. Silver Spring, Maryland: Department of Commerce.

National Research Council. 1999. "A Vision for the National Weather Service: Road Map for the Future." Washington, D.C.: National Academy of Sciences.

———. 2003. "Fair Weather: Effective Partnerships in Weather and Climate Services." Government. Washington, D.C.: National Academy of Sciences.

National Weather Service. 2005. "Working Together to Save Lives: National Weather Service Strategic Plan for 2005–2010." Government. Washington, D.C.: National Oceanic and Atmospheric Administration.

———. 2011. "The Historic Tornadoes of April 2011." Service Assessment. Silver Spring, Maryland: Department of Commerce.

———. n.d. "Eyes on the Sky: A Day in the Life of an Incident Meteorologist (IMET) on the Front Lines of a Wildfire." Government site. Weather Ready Nation. Accessed May 10, 2016. https://www.weather.gov/news/imet-article.

NWS Vision for IDSS. 2016. YouTube. National Weather Service IDSS Webinar. https://www.youtube.com/watch?v=SWbYnqrlt0A&feature=youtu.be.

Priyadharshini, Esther. 2003. "Coming Unstuck: Thinking Otherwise about 'Studying Up.'" *Anthropology & Education Quarterly* 34 (4): 420–37.

Ravetz, Jerome. 1971. *Scientific Knowledge and Its Social Problems.* Oxford: Clarendon Press.

Rosenfeld, Jeff. 2000. "Do We Need the National Weather Service?" *Scientific American*, 28–31.

Samenow, Jason. 2015. "Senate Bill Proposes Centralizing Weather Service Forecasting in 6 Regional Offices." *The Washington Post*, June 16, 2015, sec. The Capital Weather Gang. https://www.washingtonpost.com/news/capital-weather-gang/wp/2015/06/16/senate-bill-proposes-centralizing-weather-service-forecasting-into-6-regional-offices/.

Swanson-Kagan, Joanne, J. E. Ten Hoeve, III, Andrea Bleistein, J. Pavlow, J. Morrow, and C. Draggon. 2016. "Update on the NWS Operations and Workforce Analysis." Oral Presentation presented at the American Meteorological Society, New Orleans, January 11. https://ams.confex.com/ams/96Annual/webprogram/Paper287551.html.

The National Academies. 2012. "Weather Services for the Nation: Becoming Second to None." Washington, D.C.: National Research Council.

The Weather Service Modernization Act. 1992.

U.S. Department of Commerce. 2011. "NOAA Strategic Priority: Building a Weather-Ready Nation." Silver Spring, Maryland.

"Weather Forecasting: Systems Architecture Needed for National Weather Service Modernization." 1994. Report to Congressional Requesters B-255498. Washington, D.C.: United States General Accounting Office.

Winner, Langdon. 1993. "Upon Opening the Black Box and Finding It Empty: Social Constructivism and the Philosophy of Technology." *Science, Technology & Human Values* 18 (3): 362–78.

CHAPTER 6

Coping with Indonesia's Mudflow Disaster

Anto Mohsin

Introduction

A mixture of hot steam, water, and mud suddenly erupted from underground in the middle of a paddy field in Renokenongo Village in the Sidoarjo District, East Java, Indonesia on Monday morning 29 May 2006.[1] The source of the eruption was not far from an exploratory well of Lapindo Brantas, an energy company that was drilling in the area to look for natural gas deep inside the earth. What started as a small discharge quickly turned into a huge geyser. *Tempo* magazine reported that at one point, the mud discharged was as high as eight meters and it was hot enough to boil an egg (Widyanto et al. 2006a). The mud and hot gases have been spewing continuously for more than 11 years and there is no sign of stopping. Earthen dams with an average height of 10 meters and width of 15 meters now contain a vast pool of mud of about 640 hectares (Jajeli 2015).

This ongoing disaster has submerged more than a dozen villages, hundreds of houses, dozens of schools, factories, paddy fields, mosques, local grocery stores, cemeteries, a long segment of the Gempol-Surabaya toll

A. Mohsin (✉)
Northwestern University in Qatar, Doha, Qatar
e-mail: anto.mohsin@northwestern.edu

S. Amir (ed.), *The Sociotechnical Constitution of Resilience*,
https://doi.org/10.1007/978-981-10-8509-3_6

Fig. 6.1 A partially submerged high voltage electricity transmission tower. (Photo by the author)

road, and few high-voltage transmission towers (Fig. 6.1). But it is not just physical buildings and infrastructure that have disappeared or been damaged. The livelihoods, hopes, and dreams of tens of thousands of villagers have been destroyed as well. Many schoolchildren lost their school buildings and a chance to get an education. Countless farmers, factory workers, and small entrepreneurs lost their jobs and places of work. The mudflow has also ruined many communities and ripped apart the fabric of Sidoarjo societies. As one victim admits, the mud "has wrecked a social order built over generations" (Zuber 2009, 2).

The Sidoarjo mudflow is one of the longest continuing disasters in recent memory. In late May 2017, some survivors marked the eleventh anniversary of the disaster by praying and sowing flowers at a specific location on top of the earthen embankments (Suparno 2017). Many victims have been dealing with this disaster for more than a decade. The victims here include those who have lost their dwellings and needed to relocate (internally displaced persons, or IDPs) and those whose houses were not necessarily submerged but their livelihoods have been severely affected by the mudflow (indirectly affected victims, or IAVs). The latter group includes the mud-contaminated Porong River fishermen and aquaculturists as well as small- and medium-scale entrepreneurs and home industry workers whose businesses and shops used to be located along the now unusable Gempol-Surabaya toll road.

Disaster Studies literature provides two important insights to our understanding of disasters. First, there is a realization that disasters and accidents have become normal occurrences (Perrow 1999; Fortun and Frickel 2010) in our "risk society" (Beck 1992). Second, there is a growing awareness that there is no such thing as natural disasters (i.e., we can no longer blame many "natural disasters" solely on nature) since human agency is entangled with natural-based catastrophes (Steinberg 2000; Hilgartner 2007; Reilly 2009; Knowles 2014). In parallel, several studies have been produced to understand how disaster victims and survivors deal with their predicaments. There are those who tried to bring their grievances to courts and demand justice (Jasanoff 2007), through advocacy (Fortun 1998), to create a network of information using their social capital (Tasic and Amir 2016), or to participate in livelihood assistance programs (Thorburn 2009). In another study, the researchers use "psychosocial skills-based intervention" to reduce post-traumatic stresses among Yogyakarta teachers after the May 2006 earthquake (Seyle et al. 2013). These studies led some scholars to propose a conceptual framework to theorize how societies and their technological systems (i.e. sociotechnical systems) can be made to be more resilient in anticipating and dealing with disasters (e.g. Amir and Kant 2018).

This study contributes to the understanding of the sociotechnical constitution of resilience by examining how victims of a long (and seemingly unending) disaster have been dealing with their situations. It addresses the following two questions: How did the Indonesian mudflow victims cope with the disaster? What strategies did they come up with to respond to this ongoing catastrophe? The landscapes and infrastructures in Sidoarjo have been significantly transformed. But while the mud has been successfully contained, and roads, railways, and electricity transmission lines been altered and repaired, less attention has been paid to examine how people in Sidoarjo have been dealing with this unprecedented disaster quite resiliently. This chapter discusses the coping mechanisms of the Indonesian mudflow disaster victims and how they come up with and use them to deal with their deteriorating situations. It is important to note that not all victims adopted the strategies discussed in this chapter and that the strategies they employed were often modified and adapted to the changing conditions on the ground which include their constantly changing built environments, government policies and regulations, and the altering sociopolitical circumstances in their communities and in the country in general. Some of their coping mechanisms such as naming the disaster Lapindo Mudflow can be read as an act of active resistance to the dominant narrative of the disaster. Their coping strategies cannot be generalized or even transferred to other disaster victims. But their efforts to survive and bounce back clearly show that to create a more effective and resilient sociotechnical system, the *socio* aspect of the sociotechnical system must be addressed as importantly as the technical dimension.

The Government Responses

In general, the Indonesian government's responses to this disaster were slow, ambivalent, inadequate, and focused too much on infrastructure-related matters. It took the government three months after the first eruption to form the first task force to mitigate the disaster, for example. On 8 September 2006, the National Team for the Handling of Sidoarjo Mudflow Discharge or National Team for short (*Tim Nasional Penanggulangan Semburan Lumpur Sidoarjo, Timnas*) was created, headed by Basuki Hadimuljono, an engineering-trained, high-ranking bureaucrat from the Ministry of Public Works. The group was tasked to work for six months to "protect residents around the location of the disaster, maintain the basic infrastructure, and resolve the mudflow problem, while minimizing environmental damage" (Setyarso et al. 2006). Timnas' focus on engineering solutions to the mudflow problem meant that victims' issues were not of high priority. One disappointed victim from Jatirejo Village writes, "The National Team let the villagers down early on. It failed to state in clear terms that the villagers were victims of the disaster; instead it asserted that we were 'suffering losses'. This succeeded in convincing the villagers that the National Team's help would not be of much use" (Zuber 2009, 16).

The Presidential Decision No. 13/2006 that formed the National Team also decreed that Lapindo Brantas should provide the funds to complete the National Team's tasks. Indonesia's then Minister of Energy and Mineral Resources Purnomo Yusgiantoro was quoted by *Tempo* to say, "All costs covering the duties of this national team are to be borne by Lapindo Brantas" (Setyarso et al. 2006). But there was nothing in the Presidential Decision that could force Lapindo to fund the National Team other than a statement that reads "The cost needed to carry out the National Team's duties is to be charged to PT Lapindo's Budget" (Keputusan Presiden Republik Indonesia 2006). Consequently, the National Team became highly dependent on the decisions of Lapindo who was not at all serious to pay the Team's activities to mitigate the disaster.

Another government's response that produced dire consequences was the decision to channel the mud into the nearby Porong River. It was reported that in early October 2006 President Yudhoyono made a decision after a cabinet meeting to throw the mud to the Java Sea via the Porong River (Widyanto et al. 2006c). This decision cancelled Lapindo's initial plan to install a 20-km piping system that would carry the mud directly to the sea (Widyanto et al. 2006c). Instead, the company built pipes that directly discharge the mud to the Porong River. The untreated mud has been pumped and dumped into this man-made waterway ever since, affecting adversely the ecology of the river. Residents who cultivated aquacultures in the river and fishermen who had been catching fish there reported a significant decrease of their regular yields (Widyanto et al. 2006c; Fahmi 2010).

As for the National Team, President Yudhoyono later dissolved the task force after it had worked for seven months (its period was extended for one month by the president). Realizing that the crisis was greater than the team could handle, the government created a more permanent organization called the Agency for the Mitigation of Sidoarjo Mudflow (*Badan Penanggulangan Lumpur Sidoarjo, BPLS*) to mitigate both the physical and social dimensions of the catastrophe. In practice, however, BPLS mainly focused on building, strengthening, and expanding the earthen levees to contain the growing pool of mud.

The same Presidential Regulation that created BPLS also dictated how the victims should be helped. A map was drawn up to delineate which areas were devastated and which ones were not. On this map, Sidoarjo residents whose properties used to be located inside the so-called "affected area" would be "compensated" by Lapindo Brantas using what came to be known as the "cash-and-carry scheme" or receiving cash from the company for the loss of their lands and properties, provided they could produce verified land and/or property titles. Those outside of the "affected area" would be compensated by the government. In effect, the government divided the victims into two categories: those who were inside the "affected area" and those who were outside of it.

The two categories of victims created by the Presidential Regulation generated frictions and tensions among the IDPs because in most cases the IDPs outside of the "affected area" received compensation from the government more smoothly than the ones in the "affected area" (Batubara and Utomo 2012, 168–169). To make matters worse, victims who could not produce certificates of land or property ownership were offered an alternative compensation scheme that came to be called "cash-and-resettlement." Many IDPs refused this scheme because they did not want to resettle in Lapindo-built houses and also pointed out that this scheme was not mentioned in the original Presidential Regulation. As a result, their "compensation money" was delayed by Lapindo, which further frustrated residents in the "affected area."

What the Indonesian government did not expect was that the mudflow could not be stopped despite several attempts to do so. And its flow eventually inundated a larger and larger area. As a result, President Yudhoyono needed to issue additional regulations to revise the map four times to include additional bordering villages as worsening environmental conditions (poor air quality, methane gas coming out from underground, and foul-tasting well water) made living in those villages dangerous.

The Trigger Debate

Since the mudflow was reported by the media, experts from various engineering fields have been debating about what triggered the mudflow. On the one hand, there are those who support the "drilling hypothesis," which contends that Lapindo Brantas' reckless drilling activities caused the disaster. On the other hand, there are some scientists who ascribe to the "earthquake hypothesis," arguing that an earthquake in a nearby city Yogyakarta that occurred two days earlier triggered the eruption of the underground mud volcano.

The initial hot mud eruption was located not far from Lapindo Brantas's exploratory well, Banjar Panji Well-1 (*Sumur Banjar Panji-1*). Inevitably, fingers were immediately pointed at the company as the main culprit of the disaster. Imam P. Agustino, Lapindo's general manager, did not think that the mud eruption was caused by his company's drilling activities. *Tempo* quoted him as saying, "It came from a land crack due to the recent Yogyakarta earthquake. Lapindo does the drilling in accordance with the oil and gas industry standards" (Widyanto et al. 2006a). His statement was supported by Made Sutarsa, the head of the East Java's Energy and Mineral Resources Service. "The leak is due purely to a natural phenomenon, the earthquake," he said (Widyanto et al. 2006a). But Mulyo Guntoro, a geologist at the Center for Disaster Study in Surabaya's National Development University (*Universitas Pembangunan, UPN*), disagreed. He contended that the tremor felt in Sidoarjo was too small (two on the Richter scale) and that it needed to register at least five on the Richter scale to have an effect. "So it's funny if they attribute it to the quake," he said. Instead, Guntoro attributed the cause to Lapindo's unconventional drilling practice. He argued, "Drilling to a depth of more than 2 km should have been done at a slant, not straight as what Lapindo was doing now" (Widyanto et al. 2006a). Mulyono's sentiment is generally supported by many Sidoarjo's residents.

Scientists have been furiously debating the cause of the mudflow eruption. The earliest scientific paper to be published was the one written by a team of scientists led by the University of Durham geologist Richard Davies in the February 2007 issue of *GSA Today*, a journal published by the Geological Society of America. The scientists took into consideration the Yogyakarta earthquake, but eventually dismissed it as a main trigger of the mud eruption for several reasons. They wrote:

> The primary reasons for not considering an earthquake to be the trigger or contributing factor are: (a) no other mud volcano eruptions were reported in Java at the same time; (b) the earthquake preceded the eruption by two days; seismogenic liquefaction usually occurs during earthquake-induced shaking of sediment (e.g. Ambraseys 1998); (c) there are no reports of a "kick" during the earthquake or immediately afterward; and (d) sand, rather than mud, is more conducive to liquefaction due to earthquake shaking because it is a non-cohesive, granular sediment. (Davies et al. 2007, 8)

Instead, they concluded that "It is very likely that Lusi was initiated as a result of access by a high-pressure aquifer at depths in the region of 2.5–2.8 km through an open-hole section of the Banjar Panji-1 well to depths at which fractures could be initiated" (Davies et al. 2007, 9).

The "open-hole section" the scientists referred to has to do with the lack of protective casing inside Lapindo's exploratory well. The scientists believed that Lapindo's decision not to use a protective casing in the well under a certain required depth helped trigger the discharge of the mud from the high-pressure underground reservoir up to the surface. Their claim was supported by a number of other scientists including Amin Widodo, a geologist at the Sepuluh November Institute of Technology (*Institut Teknologi Sepuluh Nopember, ITS*). Additionally, Widodo pointed out other carelessness that Lapindo did, which include pulling a rig, which resulted in delaying the process of plugging the well and drilling in a heavily populated area (Widyanto et al. 2006b).

This "drilling hypothesis," however, has been challenged by other scientists, most notably by the University of Oslo geologist Adriano Mazzini and his team (Mazzini et al. 2007). But Davies was supported by the Australian geologist Mark Tingay and a few other scientists, including Indonesian petroleum engineers in a paper they published in August 2008 (Davies et al. 2008). The debate reached an interesting point when in late October 2008 at the American Association of Petroleum Geologists meeting in Cape Town, South Africa, a voting was administered among the attending scientists as to what they thought to have caused the mudflow eruption. Forty-two geologists believed the drilling activities was the sole cause. Three geologists thought that the Yogyakarta earthquake was the reason. Thirteen geologists were of the opinion that both factors contributed, while six scientists could not decide (Nature 2008). This voting has rarely if ever happened in the scientific community. Scientific controversies usually achieved "closure" using a number different mechanisms that include a loss of the credibility of the principal scientists who uphold a controversial scientific claim, a more convincing argument put forth by an influential scientist to settle a scientific debate, or even an inconclusive result for the competing hypotheses (Collins and Pinch 1998).

To help bolster their argument, scientists in each camp enrolled others (in most cases the media) to support their argument. Richard Davies, for example, has been quite vocal about his hypothesis. He has given interviews and has been quoted by many media. The supporters of the "earthquake hypothesis" gained a new ally when a documentary called *Mud Max: Investigative Documentary of the Sidoarjo Mud Volcano Disaster* was released in late 2009 in the United States. It premiered in the Asia Pacific region on 13 February 2010. It was screened at the Museum of Contemporary Art in Sydney. The movie's executive producer Chris Fong believed that the cause of the disaster remains unclear, but he thought that the "small group" of people who blame Lapindo Brantas received a lot of publicity and sought to portray the disaster in a more balanced way. Fong has been accused to be paid by Lapindo to make the movie, but both he and the director Gary Hayes denied this (Mardiyati 2010).

Lapindo Brantas, of course, disagreed with the voting results of the South Africa meeting, contending that "the data presented at the conference was unsourced and unrepresentative" (NewScientist 2008, 6). In 2009, their scientists (who are affiliated with the parent company Energi Mega Persada) published a paper with detailed drilling data arguing that the drilling activities could not have caused the eruption (Sawolo et al. 2009). Davies and his colleagues disputed this charge in a rebuttal paper (Davies et al. 2010), which was followed by the authors' response to this paper (Sawolo et al. 2010).

The scientific debate continued. In 2013, another team of scientists published a paper arguing that the Yogyakarta earthquake triggered the eruption (Lupi et al. 2013). This paper was accompanied by a scientist's commentary that states:

> It remains unclear why waves from previous earthquake which also would have experienced focusing by the curved layer of shale, did not trigger a mud eruption. Perhaps it was the combination of drilling and the earthquake that triggered the 2006 eruption, and neither trigger alone should take the full blame. The Lusi eruption was probably inevitable, and perhaps perturbations from the earthquake and/or drilling advanced the clock. Indeed, we may never know what the final trigger was, whether it would have happened anyway, nor even if an early trigger averted a greater disaster, had pressures continued to build up. (Davis 2013, 593)

The latest paper on the subject was written by several familiar scientists (Tingay et al. 2015) who argue against the earthquake hypothesis advanced by Lupi and others. The lead author in an interview was quoted "We're now 99 percent confident that the drilling hypothesis is valid" (Nuwer 2015). But other scientists expressed a doubt. One of them, Stephen Miller, is quoted to say "I'm surprised that the authors could arrive at such a strong conclusion from such inconclusive data. All science screams that Lusi is natural" (Nuwer 2015). This so-called trigger debate is still going on (Drake 2016).

Making the Best Out of Bad Situations

In the days and weeks following the first discharge of the mudflow, many residents who had to evacuate their houses went to the New Porong Market (*Pasar Baru Porong*). This was a new installation built to accommodate the expanding number of retailers in the area. Many displaced residents quickly made this market as their make-shift refugee camp. There were 282 kiosks, 51 shops, and 146 sanitation facilities at the market that were made available to refugees as temporary shelters (Hadimuljono 2012, 132). By late January 2007, it was reported that close to 15,000 IDPs were staying at the camp (Muhtada 2008). They were mainly the less well-off Sidoarjo residents who did not possess documents proving their land and property ownership. In this regard, Schiller et al. point out that these IDPs were not just homeless, but in fact also "stateless," that is, unrecognizable by the state as being legitimate property owners (Schiller et al. 2008, 54). Those who could afford it moved elsewhere to stay with family members or to rent a place. Some 700 IDPs were accommodated at the Renokenongo Village Meeting Hall (*Balai Desa Renokenongo*) (Akbar 2007, 100; Hadimuljono 2012, 131). For many victims staying at these two shelters provided a temporary relief to their predicaments.

It was not always easy staying at the Porong New Market, however. Families who were staying there lived in small and cramped places with little or no privacy. An Airlangga University psychiatrist who examined the refugees noted that many of them were depressed (Muhtada 2008). They got angry easily and were prone to getting into conflict with each other (Dhyatmika 2007; Zuber 2009). Consequently, social tensions ran high in the camp (Muhtada 2008). Many suffered from depression and were called "*bocor alus*," a term repair technicians use for a slowly leaking tire. "They look like they are fine, but then one day, they [could] end up in a mental hospital," said one refugee (Dhyatmika 2007).

Despite these drawbacks, many IDPs were determined to make the best out of the uncomfortable situations. Some residents, for example, decided to build their own kindergarten at the camp. Calling it the Muhajirin Kindergarten, the founders of this school recruited some parents as teachers.[2] They obtained school supplies from various donations. A few residents even volunteered to create the chairs and tables for this school. Having a place to go to school at the "camp" was a welcome respite both to the kindergarteners and their parents. They were sheltered, so to speak, from going to other schools and being bullied and called various names, such as "children of the mud" (Korbanlumpur 2008). At one point, two young foreign tourists named Eef and Lowie who were backpacking in Indonesia visited the kindergarten, bought some supplies for the kids, stayed, and interacted with them for a few days (Korbanlumpur 2008). It wasn't just kindergarteners who found a chance to play and learn at the market. Students from the Asy Syadzili Islamic Boarding School briefly studied here, too (Zuber 2009, 45).

Other refugees took up various other activities to cope. Elis Susanti, for example, decided to paint as a therapy (Dhyatmika 2007). To make ends meet, many looked for new jobs or started to sell foods and snacks. Others worked as laborers or parking attendants. Some others became *ojek* drivers using their motorcycles to drive "tourists" who wanted to look at the expanding mud pool. These *ojek* motorcyclists also tried to sell Video Compact Discs (VCDs) of the disaster area to those who hired their service. Schoolchildren were encouraged to write. Nur Ainiyah, an elementary school student, for example, wrote that she could not sleep well at the camp because there were too many mosquitos and that she yearned to stay at her own house so she could sleep better and continue her study (Aniyah 2010). Some teenagers were trained by volunteer activists to write personal blogs to express their thoughts about what they and other victims were going through. One blogger named Daris Ilma was inspired by the activity that she changed her major in college from physics to communication. A few others conveyed their emotions in a documentary called *Yang Tak Ingin Terbenam*(*Those Who Do Want to be Submerged*) produced by DAAI TV.

A few people took the initiative to create a community radio station called *Radio Komunitas Suara Porong* 107.8 FM. One of its broadcasters, Lilik Kaminah, told me the genesis of its founding. It started with a lack of and inconsistent information the refugees received at the market. Some refugees then decided to start a radio station. Through a volunteer they were put in touch with a private radio station and some crews from KBR 68H Jakarta came to Sidoarjo and helped create this community-based radio station. The crews trained some people to collect stories and broadcast them. The broadcasters aired situations that were happening in and around the market and disseminated other information to the community. They worked voluntarily and relayed information using short messaging services (SMS). Sometimes, they would broadcast an impromptu event from the field as it was unfolding on the ground. When this occurred a reporter would broadcast the story by speaking to his or her mobile phone while at the station a colleague would hold his or her phone to the microphone. But the station sometimes stopped broadcasting because the equipment broke or no broadcaster was available. At the market, some refugees used the radio station to vent their frustrations. They took the mic and expressed their angers using many expletives and the radio broadcasters let them do that (L. Kaminah, personal communication, December 12, 2016).

Lumpur Lapindo

One controversial aspect of the mudflow is the naming of the disaster. The majority of the victims (including all of whom I spoke with in December 2016 when I visited Sidoarjo) have been persistently calling this disaster *Lumpur Lapindo* (Lapindo Mudflow). They believe that the company Lapindo Brantas is responsible for the disaster and must be held accountable despite the ruling of the Indonesian Supreme Court that absolved the company in a decision issued in 2009 (Davis 2013; Mardiyati 2010). Several factors contributed to this belief. First, many victims like to point out that prior to Lapindo's exploratory well, there was no mudflow at all. It was only after the Banjar Panji-1 was drilled that the mud started to come out from underground. Although the company has denied any wrongdoing, the mud discharge location was close to the borehole and there have been reports that the company was negligent in using a protective casing in the well. Second, as indicated earlier, more independent scientists (i.e., those who are not affiliated with Lapindo Brantas) have argued for the "drilling hypothesis," lending further support to the notion that the company was to blame for the disaster. Third, President Yudhoyono has issued a regulation mandating the company to pay for the cost of the mudflow mitigation and compensation of some of the victims, to which Lapindo has agreed to do. All of these have convinced many victims that Lapindo was accountable for the catastrophe (Schiller et al. 2008).

Calling it *Lumpur Lapindo* is in contrast to another name for this disaster, which is *Lumpur Sidoarjo* (Sidoarjo Mudflow) or Lusi for a contraction of the term. The majority of scientists who have analysed and written about the mudflow tend to use this designation (for e.g., Davies et al. 2007; Mazzini et al. 2007; Tingay et al. 2008; Istadi et al. 2009; Lupi et al. 2013). Karni Ilyas, the host of one of Indonesia's famous TV shows *Indonesia Lawyers Club*, on an episode called "Presiden Tagih Utang Lapindo" (President Demands Lapindo's Debt) chooses to also use this term, arguing that whenever a disaster occurs in the world, usually it is called by the name of the region in which it occurs such as the Bhopal disaster, the Alaskan [Oil Spill], the Chernobyl disaster, and the Buyat Bay pollution.[3] The editor-in-chief of *Surabaya Post* Dhimam Abror Djuraid opined that the term *Lumpur Lapindo* focuses the responsibility on Lapindo Brantas, ignoring the government's role in disaster mitigation (Novenanto 2010b). It was this rationale that led Djuraid to use the term Lumpur Porong (Porong Mudflow) in his newspaper, after the name of a sub-district in Sidoarjo where the mudflow discharge is located (Novenanto 2010b). Articles that appear in international newspapers (e.g., *New York Times*) and journals (e.g., *Nature*) use the more neutral phrase *LumpurSidoarjo*. Additionally, the government agency created to contain and mitigate the mudflow is called *Badan Penanggulangan Lumpur Sidoarjo* (Agency for the Mitigation of Sidoarjo Mudflow).

But the victims are not alone in choosing to call the catastrophe *Lumpur Lapindo*. The government of the District of Sidoarjo does not like the name Lumpur Sidoarjo, "since that name contributes to the image of the entire district of Sidoarjo as a dangerous place, an impression that has already frightened away investors, tourists, and trade" (Schiller et al. 2008, 56). Others who use this term such as Hendrik Siregar, an activist with the Mining Advocacy Network (*Jaringan Advokasi Tambang, Jatam*), argues that calling it *Lumpur Lapindo* helps fight systematic efforts to call this disaster a natural disaster rather than an oil and gas exploration accident (JPNN.com 2010). Many civil society organizations prefer to use this name too. When in mid-2008 a media advocacy team was formed to write and publish a newsletter about the plight of the victims, it named its publication *Kanal Saluran Aspirasi Korban Lapindo*. As we will see later, the creation of the *Kanal Newsroom* provided a space for some survivors to tell their stories.

The decision to call the disaster the Lapindo Mudflow was one of the earliest and effective coping mechanisms the victims adopted. It was a strategy that served as an act of active resistance as well. To many victims, calling the mudflow *Lumpur Lapindo* was to counter the "natural disaster" narrative that was put forth by Lapindo and some government officials. The term is politically charged and reduces the blame to one entity. But in the face of attempts by Lapindo to shape public opinion by claiming that the catastrophe was a natural disaster (Novenanto 2010a), this seemingly simple strategy proved quite effective in fighting it. It allowed them to raise public awareness of the disaster and enroll others to their cause. Perhaps more important, calling it Lumpur Lapindo denaturalizes the "natural" aspect of the disaster and stresses the implication of human agency and corporate decisions in this catastrophe, supporting one important insight of disaster studies scholarship (Steinberg 2000; Hilgartner 2007; Reilly 2009; Knowles 2014).

In many subsequent articles written and stories narrated by the victims, they have consistently been using the term Lapindo Mudflow to call the tragedy that befell them. Not only does the term invoke the name of the company many think should be held responsible for the disaster, it also allows people to direct their anger and frustrations to the owner of Lapindo, Aburizal Bakrie. An effigy of Bakrie was made, paraded, and thrown into the pool of mud during the seventh and ninth anniversary commemorations of the disaster.

Victim Groups

The names of a few of the groups formed by some of the victims also bear the term. Residents of the former Renokenongo Village, for example, created a group called *Paguyuban Warga Renokenongo Korban Lapindo* or *Pagar Rekorlap* (Association of Renokenongo Residents of Lapindo Victims). There were about 563 families who became members of this group and all of them lived in the New Porong Market. Initially, this group's name was called *Paguyuban Warga Renokenongo Menolak Kontrak* or *Pagar Rekontrak* (Association of Renokenongo Residents Refusing Contract). They rejected Lapindo Brantas' lump sum payment of Rp 5 million to rent places in new areas, living expenses of Rp 300,000/person/month (about US$30/person/month), and a one-time relocation fee of Rp 500,000, arguing that doing so would break apart their community. Renokenongo villagers preferred to stay together united and to relocate as one community. The only way to do that was to receive 100 percent compensation for their lands and properties from Lapindo. They therefore vehemently disagreed with the Lapindo's compensation scheme paid with a down payment of 20 percent and in installments for the remaining 80 percent in the beginning. Consequently, they refused to evacuate from the New Porong Market to move elsewhere. After a long and drawn-out negotiation, however, they finally agreed to accept the compensation scheme (Novenanto 2009).

Pagar Rekorlap was not the only group formed by IDPs. There were a few others and each group was created for specific purposes. The largest of them was called *Gabungan Korban Lumpur Lapindo* or *GKLL* (Joint Lapindo Mudflow Victims), which consisted of IDPs from Jatirejo, Renokenongo, Kedungbendo, and Siring villages. Its members were varied and they accepted both of the cash-and-carry and cash-and-resettlement schemes (Novenanto 2009). Not all IDPs agreed to the latter scheme, however. Those GKLL members who voiced their disagreement formed a different association called *Gabungan Penduduk Pendukung Peraturan* President or Geppres (Joint Residents In Support of the President Regulation) who contended that there was nothing in the President Regulation No. 14/2007 that allows Lapindo Brantas to offer a cash and resettlement scheme and force IDPs to move to the company's newly built housing complex called Kahuripan Nirwana Village (KNV). This scheme created a lot of angst among many victims because the price of a KNV house is higher than the cash they would receive and that they would need to pay the difference using their own money. In other words, instead of compensating the victims for the loss of their lands and properties, Lapindo tried to sell them their newly built properties. But there were some victims who accepted the cash-and-resettlement scheme and they split into another group called *Tim 16 Perumtas* (16 Perumtas Team) (Lay 2017; Novenanto 2009).

Perhaps the most politically active group was the *Korban Lapindo Menggugat* or *KLM* (Lapindo Victims Suing), which was a group of IDPs that formed following a dam breach in Gempolsari Village in December 2010 (Lay 2017). In addition to demanding compensation from Lapindo, KLM argued against further drilling in Sidoarjo and called BPLS to reinforce the northern part of the levees (Lay 2017). Their demands, however, were largely ignored. Then the group partnered with a non-governmental organization, Urban Poor Consortium (UPC), and changed their approaches. With UPC's help, KLM managed to obtain new birth certificates for 400 of its members. Having a birth certificate is important to claim their rights as citizens, including compensation money for the disaster. As Cornelius Lay writes, this win was "an important symbolic victory" for them (Lay 2017, 41).

In the 2014 election season, KLM members became active volunteers in support of Joko Widodo's presidential candidacy. Joining forces with another community association in Surabaya's Sren River and three organizations—*Jaringan Rakyat Miskin Kota* or JRMK (Network of the Urban Poor), Urban Poor Linkage or UPLINK, and *Jaringan Rakyat Miskin Indonesia* or *Jerami* (Network of Indonesian Poor People)—KLM made a pact with one of the two presidential candidates on 29 May 2014. This transformed KLM from a victim advocacy group into a political volunteer organization to help Joko Widodo (better known by his nickname Jokowi) win the election (Lay 2017). Jokowi, an outsider candidate who served as the Governor of Jakarta and the mayor of Surakarta beforehand, benefited greatly from volunteers. Although Jokowi was picked by one of Indonesia's big political parties as their presidential candidate, their selection was last minute and the party bosses did not trust him completely. Consequently, Jokowi came to rely more on his "grassroots volunteers than to machine politicians" to campaign on his behalf for the election (Mietzner 2014, 112). KLM members collected coins for Jokowi's campaign fund, visited and convinced residents to vote for Jokowi, "rapped" a campaign song, distributed flyers, placed posters and banners around the earthen levees, and served as witnesses at the polling stations (Lay 2017, 50–55). Their support for Jokowi paid off when the new president decided in 2015 to provide bailout funds for many victims.

Annual Commemorations of the Disaster

The political contract made between KLM and Jokowi occurred at an opportune moment. He came to the disaster site when victims flocked there to commemorate the eighth year of the disaster in May 2014. He was therefore able to meet many people. The earliest record in the media of this yearly commemorative event took place on 29 May 2009, on the third anniversary of the disaster. Thousands of victims gathered at one location on top of the earthen embankment to remember the day when a mud discharge began and grew to engulf their entire villages (Tejo 2009). The following year, dozens of former Jatirejo Village residents and Sidoarjo artists held a three-day festival remembering the fourth year of the disaster. They did a number of activities including a performance at a "people's theatre," holding a few workshops (T-shirt printing, woodcuts, etc.), and carrying banners voicing their condemnations of the government and Lapindo as well as expressing their anger and frustrations (Kabarsidoarjo 2010).

On 29 May 2011, hundreds marched carrying posters conveying their demands: "*Adili Lapindo, Tuntaskan Ganti Rugi Korban Lumpur* (Bring Justice to Lapindo, Complete the Compensation for the Mudflow Victims)" and "*Tutup Semburan Lumpur* (Close the Mudflow Discharge)" (Kompas 2011), voicing their conditions: "*Tambak Kami Jadi Rusak* (Our Aquacultures Were Damaged)," and expressing their disappointment: "5 Tahun Tidak Tuntas, Presiden Gagal (5 Years Unresolved, the President Failed)" (Kompas 2011).

The victims' plight drew sympathies from people outside of Sidoarjo. On the fifth anniversary on the mudflow, for example, a few people blogged about the disaster. Herawanto Samad concisely expressed the unfortunate conditions of many victims and called for the government to take a more serious effort to solve it. He wrote, "The government must create a clearer regulation, especially who is to blame for the Lapindo Mudflow so that no people would be the victims of a greediness of one group looking to reap profits from the wealth contained under this earth" (Samad 2011). Samad blogged on kompasiana.com, a free blogging platform provided by *Kompas*, one of the big media companies in Indonesia. Usually, a popular kompasiana.com posting would be highly referenced and returned as one of the top search results on Google. Two additional kompasiana.com bloggers also wrote about the fifth anniversary of the mudflow, reminding the public of the ongoing disaster in the country (Laksana 2011; Subagyo 2011).

In the following year, some of the IDPs, particularly those whose houses were in the affected area, decided to commemorate the disaster by holding a collective prayer (*istighosah*) asking the Almighty for help with their difficult problems. Holding a communal prayer is both a social and religious coping strategy. The found a strength in number and in their faith. In the face of uncertainties, a powerful and well-connected corporation, a seemingly uncaring government, and an unreliable justice system, many victims believed that a Supreme Being would listen and hopefully answer their prayers if they did so collectively and persistently. For two nights they gathered to pray that government leaders and Lapindo executives would fulfil their promises to compensate them. They have grown tired and frustrated because their requests have fallen on deaf ears. By this time and according to an agreed-upon schedule, all of the residents of the "affected area" should have received their money from Lapindo. But many were still waiting for the 80 percent remainder of the promised fund that Lapindo said would be disbursed in installments (Andriansyah 2012).

In addition to praying, some mudflow victims did a number of other activities on 29 May 2012. One that caught the attention of the media was to revive an old ritual in response to a disaster. In one of Indonesia's Hindu-infused legends, a hideous giant called Betara Kala comes to a village wreaking havoc on the community by creating death and destruction. One effective method to get rid of the giant was by hitting a *kentongan*, a traditional communication device made out of hollowed-out wood. About 80 residents composed of men, women, and children hit *kentongans* to symbolically get rid of Aburizal Bakrie who they deem as their Batara Kala. One participant named Gugun even made the connection explicitly. He was quoted as saying, "Aburizal Bakrie is identical with Betara Kala. Let's rid of him from Sidoarjo by hitting these *kentongans*" (Mawuntyas 2012).

A 2.5-meter effigy of Aburizal Bakrie, as a patriarch of the Bakrie Group that owns Lapindo Brantas and deemed to be the single most responsible person for the disaster, was carried in a procession and then dumped into the mud pool by victims who participated in the seventh commemoration event in late May 2013. Abdul Rokhim, the coordinator of the event is quoted as saying, "He [referring to Bakrie] must be responsible for the mudflow. But in these past seven years he seems to be undisturbed [by our repeated requests for full compensation payment]. But, we will continue to contest, our voices never die" (Putro 2013). At the event, three victims used their bodies to make statements. They jumped into the pool of mud and swam while journalists and other media professionals recorded the event.[4] All of this was done to express their prolonged exasperations of broken promises by Lapindo and of being largely neglected by the central government in the wake of a continuing disaster.

Another effigy of Bakrie, this time of him being handcuffed, was made by about twenty volunteer IDPs for about one month on the eve of the ninth anniversary of the mudflow in 2015. The much bigger effigy (about 5 meters tall) was paraded for about 1 km in a procession that involved 100 people (Tribunnews 2015).[5] The commemoration was also marked by an event the survivors called the Pulang Kampung Festival (*Homecoming Festival*). They built 11 huts on top of the dried mud to represent the condition of their village lives prior to the disasters. Each of the huts offered Indonesia's traditional dishes for the participants (Kompas 2015). The main purpose of the festival was to reminisce their lives as one community in their disappeared villages. After the mudflow, many former residents had to relocate to different places, tearing apart the social fabric that they have built for years. The festival was supposed to remind of them their existence as a community. Neighbours came together to the event to meet in the hope of rebuilding their broken social ties (Riski 2015). As before, they held a collective prayer during this event (Astuti 2015).

A modest monument in a shape of a pentagon was erected and placed on 29 May 2014 on the eighth anniversary of the catastrophe. The simple structure is painted white with the following words (translated into English) written in black (Fig. 6.2): The Lapindo Mudflow Has Drowned Our Villages/Lapindo Only Makes Empty Promises/The State is Neglectful in Rebuilding Our Lives/Our Voices Will Never Die/So This Nation Will Remember/29 May 2014.

Also on 29 May 2014, there were statues placed on the dried surface of the mud pool. The sculptor Dadang Christanto created 110 statues all with their arms held up and palms opened indicating that they have been "waiting" to receive their relief funds. Three days before the commemorative event to take place in Sidoarjo in 2014, the statues were placed in the

Fig. 6.2 The Lapindo Mudflow tragedy monument. (Photo by the author)

pool of mud not far from where the monument to be located.[6] The statues were made of cement that was mixed with mud and other materials. Volunteers carried those statues to a specific location of the earthen levees called Point 21 to be placed on top of a dried surface of the mud pool. According to the artist, they symbolize the disaster survivors' resilience as victims who had lost many things in the disaster but still could survive (Surya 2014). Because the mud continues to discharge, 2.5 years later, a large part of the statues have been drowned. In December 2016, all that could be seen was their shoulders and heads (Fig. 6.3).

These statues and the monument became visible artefacts of the victims' attempt to remind everyone of their sufferings and to fight the nation's memory loss as less and less attention is being paid to this still ongoing disaster. As time goes on, less reporting is being done about the disaster except on the eve and during the annual commemoration. This is one reason why the residents continue to hold the annual commemoration. Another reason is to attract allies to support their cause. The annual remembrances provide a space for symphatizers to participate and help survivors make the events more meaningful to them.

Fig. 6.3 Dadang Christanto's submerged statues. (Photo by the author)

It was most likely because of this reason that Christanto participated and contributed again in the tenth anniversary commemoration in 2016. This time his artwork consisted of 300 bamboo sticks, each is attached with a piece of used fabric owned by the survivors. Calling the installation "*Gombal*" (Rags), he said that the victims "identify themselves as *gombal*, marginalized and uncared for. This is a picture of a government who still treats its citizens as *gombal*" (Riski 2016). Christanto also said that the word *gombal* has another meaning, which is "nonsense" or "broken promise" symbolizing the empty pledges that many government officials and politicians have made to many victims. "This Lapindo Mudflow is the most extreme and monumental in terms of people's sufferings. It is really severe," he said (Riski 2016).

The annual commemorations have served more than just yearly remembrances of the disaster. They helped survivors cope with negative consequences of the disaster collectively as a community. Once a year they come together to reminisce their lives before the disaster, pray for things to get better, and vent their angers and frustrations. They have even used the occasion to fight a new battle. In early 2016, there was news that Lapindo Brantas would start a new well in the area. This was met with fierce resistance from the residents there. But the company has been adamant to carry out its plan and the survivors used the tenth anniversary event to resist this new drilling effort. If there was one lesson to be learned from this disaster it is that the government "should not give mining permit in a heavily populated area," said Rere Christanto of Indonesia's non-government environmental organization *Wahana Lingkungan Hidup* or *Walhi* (Environmental Forum) and one of the disaster's survivors (Riski 2016).

Kanal Newsroom

In mid-July 2008, 42 members of various NGOs and 19 representatives of victims met at Syahida Inn in Ciputat, Jakarta. After listening to the victims' stories, the roles played by the NGOs to help them, and analysing the problems that all of them were facing on the ground, the meeting concluded with several recommendations, one of which was to communicate better among the groups and to communicate the challenges and struggles faced by the victims to an outside audience. Consequently, a new media advocacy team called the Kanal Newsroom was born to collect and report stories of the plight of the victims. The Kanal Newsroom was not a formal news organization, but the people involved in it were supported financially and logistically by the AirPutih Foundation. They also decided to expand an existing website: www.korbanlumpur.info to post stories from the field (Hamdi n.d.).

Several young activists were asked to spearhead the efforts and all of them agreed to do so part time. Mujtaba Hamdi of *Lapis Budaya Indonesia* headed the editorial team of a newsletter they called *Kanal* (Canal) with a slogan "*Saluran Aspirasi Korban Lapindo*" (The Channel of Lapindo Victims' Aspirations) under its masthead. As the name and slogan suggest, the newsletter would be published with an intention to channel the mudflow victims' stories so the publics across the country know the challenges and sufferings they were facing daily. In an unpublished final report written for the Lapis Budaya Indonesia Foundation, Hamdi wrote that survivors and other Sidoarjo residents sometimes visited their Kanal Newsroom "Secretariat" in Sidoarjo to talk, get information, or just to lament about their conditions. They recalled their lives before the mudflow, the havoc it wreaked, and their attempts to get their rightful compensation. Their stories and input served as fodder for the stories featured on the Kanal bulletins and website. Between July 2008 and February 2009, the Kanal team produced more than 190 written pieces. Kanal Newsroom also setup and operated a small radio station *KanalRadio*, although in its development it faced more challenges than it could handle (Hamdi n.d.).

Another bulletin called *Somasi* also came into existence around the same time that the Kanal newsletter was published. Supported by the Tifa Foundation, the editor-in-chief was Yayan Sakti Sundaru. The Tifa Foundation cooperated with the Lembaga Penelitian dan Pengabdian Masyarakat (Research and Service Agency) of the Airlangga University, which at the time was headed by Dr. Bambang Sektiari Lukiswanto. Both the *Kanal* newsletters and the *Somasi* bulletins became "alternative media" that managed to raise important issues of victimhood, the victims' resistance, and "the commodification of disaster" (Novenanto 2010a).

The *Kanal* bulletins were published in 10 editions between August 2008 and February 2009 and each edition was printed 1000 copies to be distributed in 14 areas in Sidoarjo. The initial intention was to create the newsletter to inform residents of their legal and civic rights as well as an education medium for them to fight for their rights. But as it evolved, the newsletters and the website became important sources for the media, researchers, and even politicians who campaign for a local office. Volunteers from the Australian-Sidoarjo Assistance Project (ASAP) helped translate some articles into English, reaching a wider audience (Hamdi n.d.).

What is important to note about the *Kanal* bulletin and the website korbanlumpur.info is that it brought about three positive unintended consequences. First, some victims or children of victims became involved in the writing of stories for the newsletter and website. It started with a small training workshop in which several youths learned how to write, take photos, and post blogs. One of the youngsters who participated in this effort even discovered a newfound passion in journalism. Daris Ilma, a daughter of Cak Rokhim, was initially enrolled in Airlangga University as a physics student, but then switched to another university to study communication (D. Ilma, personal communication, December 13, 2016). Another writer and contributor Rere Christanto became active in Walhi and he especially remembered about the piece he wrote on the deteriorating but neglected health conditions of many of the victims for one of the *Kanal* editions (R. Christanto, personal communication, December 12, 2016). Second, another good outcome was a wider support that the victims started to receive following the publication of their stories on the website. Hamdi writes in his report that the website became a campaign platform on behalf of the victims for anyone who is interested in studying the mudflow, although I have to add that since the website is written in Indonesian, it has a limited reach. Third, there was an increased awareness among the victims in demanding their right to information, especially related to the compensation money that they were entitled for (Hamdi n.d.).

Another significant publication that *Kanal* did was to collect 29 stories and publish them in a book titled *29 Cerita Menentang Bungkam* (29 Stories Fighting Silence) edited by Mujtaba Hamdi. Based on the premise that Lapindo has been using its power and capital to silence victims' voices in the mainstream media, the book was written to fight Lapindo's effort. It was not just Lapindo that the contributors to the edited volume tried to fight, but also the Indonesian police, the Indonesian justice system, and the Indonesian government who seemed incapable of standing up to Lapindo (Hamdi 2010). The number "29" signifies the date of the mudflow eruption in May 2006. Eight writers (at least four of them were victims of the mudflow) interviewed and wrote several pieces for the anthology. Divided into seven sections (economy, social, culture, education, environment, law, and politics), the eight writers collected victims' stories that were then written up and placed in these eight categories. All but the "environment" section has four stories about victims who recounted their lives before the mudflow, the struggles they have been going through, their efforts to deal with their new conditions, and their hopes for building a better future after receiving their full compensation money. There is one underlying theme connecting the 29 stories. Implicit in their narratives is their gratefulness of having a chance to voice their plights. The book gave them a voice to express their circumstances, anger, frustrations, ideas, condemnations, and yearning to rebuild a life shattered by an unexpected and unexpectedly long disaster.

Conclusion

Many disaster victims of the Indonesian mudflow disaster were members of vulnerable groups but proved quite resilient. They adopted and used various coping strategies and took a number of different initiatives to deal with negative consequences of a long disaster. They marshalled resources that existed in their social, organizational, and institutional spheres; used alternative media to voice their predicaments; and worked with sympathizers and allies toward achievable goals. In doing so they interacted with the material objects in their surroundings to tell their stories and fight memory loss, used language to resist a dominant narrative of the disaster, and manuevered within the sociopolitical, legal, and economic constraints to achieve a goal of being compensated fairly.

When the mud started flowing into their villages, many people immediately used what was available to them as a makeshift refugee camp. In this "camp," they did their best to make the best out of miserable conditions. Many suffered depression and other mental illnesses as reported by the media. But there were others who decided to build a kindergarten, held classes, and started a community radio. Many IDPs eventually moved out of the New Porong Market after staying for up to three years there.

Despite the controversies surrounding the causes of disaster and notwithstanding the unsettled scientific debate about what triggered the eruption, almost all of victims associated this disaster with Lapindo Brantas, the energy company publicly deemed responsible. They have been calling this disaster the Lapindo Mudflow Disaster pointing to the human aspect of and calling attention to the high degree of human agency involved in contributing to this catastrophe. Doing so enabled victims to actively resist the dominant narrative about the disaster told by Lapindo Brantas.

Calling this disaster by another name would not be effective without taking part in creating different associations. Forming different victim groups not only helped the victims collectively fight for what they wanted, but also allowed some of them to get politically involved. What one group (KLM) did by becoming a volunteer group for one of Indonesia's presidential candidates in 2014 was a response to a changing political climate as well as to transforming ecological and sociocultural systems. A few of them even became candidates themselves in a local election (Christanto 2010a). Others decided not to exercise their right to vote (Christanto 2010b). In short, becoming involved politically allowed them to deal with their changing landscapes and circumstances in the wake of the disaster.

Remembering the disaster every year was another coping mechanism that disaster victims employed. These events have managed to keep the attention of the national media and helped enroll others to their cause. Commemorations and monuments left behind have been serving as a reminder of how this continuing devastating disaster are still in the minds of many villagers who have lost so much but still received little attention from the government. They also act as a way to anticipate potential future disaster by calling attention to their neglected plight as a warning to other people in the country to demand accountability from the government.

A media advocacy team Kanal Newsroom facilitated disaster victims to express their own voices, showing that this coping strategy helped them fight attempts to silence or drown out their voices in the mainstream media. Their stories helped outsiders learn more about their plights and bring attention to unsettled problems that they still face until today. Organizing and disseminating information this way allowed the victims to coordinate, organize, and plan actions to achieve their objectives of getting compensation and rebuild their lives.

Although we cannot treat the Indonesian mudflow victims as a homogeneous group or generalize their coping mechanisms, their efforts to deal with and recover from a disaster point to what Sulfikar Amir and Vivek Kant (2018) refer to as sociotechnical resilience. It illustrates how their endeavors to bounce back from a disaster were founded in the idea that both the sociopolitical and the technical factors were entangled and needed to be employed to cope and survive a long disaster. Since this disaster is ongoing, this is not the end of their stories. Both the IDPs and IAVs are still trying to deal with some lingering environmental, health, and social issues in the area. Further studies of how they continue to deal with these persistent problems are needed.

Notes

1. See Hadimuljono pages 80–81 for before and after aerial photos of the location near the center of the mudflow eruption.
2. Muhajirin means people who migrate from one place to another.
3. The TV show can be viewed in segments on https://hotmudflow.wordpress.com/ (accessed 15 August 2017).
4. A video of this can be seen online at http://korbanlumpur.info/2013/05/warga-arak-ogoh-ogoh-aburizal-bakrie/
5. Video of the Ninth Year Commemoration can be viewed online: http://m.tribunnews.com/video/2015/05/29/newsvideo-peringati-9-tahun-lumpur-lapindo-warga-arak-ogoh-ogoh
6. www.merdeka.com/foto/peristiwa/373325/20140528115616-pameran-patung-sambut-peringatan-8-tahun-bencana-lumpur-lapindo-006-nfi.html

References

Akbar, Ali Azhar. (2007). *Konspirasi Di Balik Lumpur Lapindo Dari Aktor hingga Strategi Kotor*. Yogyakarta: Penerbit Galangpres.

Ambraseys, N.N. (1998). Engineering Seismology: Earthquake Engineering and Structural Dynamics. 17, 1–105.

Amir, S. and Kant, V. (2018). "Sociotechnical Resilience: A Preliminary Concept." *Risk Analysis*, 38: 8–16.

Andriansyah, Moch. (2012, May 29). "6 Tahun lumpur Lapindo, warga gelar istighosah." https://www.merdeka.com/peristiwa/6-tahun-lumpur-lapindo-warga-gelar-istiqhosah.html (accessed August 25, 2017).

Aniyah, Nur. (2010, March 13). "Di Desaku Dan Paspor." http://korbanlumpur.info/2010/03/di-desaku-dan-paspor/ (accessed August 25, 2017).

Astuti, Runik Sri. (2015, May 29). "Warga Peringati 9 Tahun Semburan Lumpur." Kompas.

Batubara, Bosman and Paring Waluyo Utomo. (2012). *Kronik Lumpur Lapindo Skandal Bencana Industri Pengeboran Migas di Sidoarjo.* Yogyakarta: INSISTPress.

Beck, Ulrich. (1992). *Risk Society Towards a New Modernity.* London, Thousand Oaks, New Delhi: Sage Publications.

Christanto, Rere. (2010a). "Menanyakan Kembali Teriakan Dari Dalam Itu." In Mujtaba Hamdi (editor) *29 Cerita Menentang Bungkam Aneka Suara Dan Tuturan Korban Lumpur Lapindo.* Penerbit Kanal.

Christanto, Rere. (2010b). "Aku Memilih Untuk Tidak Memilih." In Mujtaba Hamdi (editor) *29 Cerita Menentang Bungkam Aneka Suara Dan Tuturan Korban Lumpur Lapindo.* Penerbit Kanal.

Collins, Harry and Trevor Pinch. (1998). *The Golem, Second Edition.* Cambridge, UK: Cambridge University Press.

Davies, Richard J., Richard E. Swarbrick, Robert J. Evans, Mads Huuse. (2007). "Birth of a mud volcano: East Java, 29 May 2006." *GSA Today,* 17(2): 4–9.

Davies, Richard J., Maria Brumm, Michael Manga, Rudi Rubiandini, Richard Swarbrick, Mark Tingay. (2008). "The East Java mud volcano (2006 to present): An earthquake or drilling trigger?" *Earth and Planetary Science Letters* 272: 627–638.

Davies, R., Michael Manga, Mark Tingay, Susila Lusianga, Richard Swarbrick. 2010. "Sawolo et al. (2009) the Lusi mud volcano controversy: Was it caused by drilling?" *Marine and Petroleum Geology,* 27: 1651–1657.

Davis, Paul. (2013). "Triggered mud eruption?" *Nature Geoscience,* 6: 592–593.

Dhyatmika, Wahyu. (2007, June 4). "Progress…What Progress?" *Tempo English Edition.*

Drake, Phillip. (2016). "Multiple visions of Indonesia's mud volcano: understanding representations of disaster across discursive settings." *Disasters,* 40(2): 346–364.

Fahmi, Miftakhul. 2010. "Bandeng Tinggal Ikon, Udang Hanya Lambang." In Mujtaba Hamdi (editor) *29 Cerita Menentang Bungkam Aneka Suara Dan Tuturan Korban Lumpur Lapindo.* Penerbit Kanal.

Fortun, Kim. (1998). "The Bhopal Disaster: Advocacy and Expertise." *Science as Culture.* 7(2): 193–216.

Fortun, Kim, Scott Frickel. (2010). "Making a case for disaster science and technology studies" https://fukushimaforum.wordpress.com/online-forum-2/online-forum/making-a-case-for-disaster-science-and-technology-studies/ (accessed August 25, 2017).

Hadimuljono, M. Basuki. (2012). *Semburan Lumpur Panas Sidoarjo.* Spirit Komunika.
Hamdi, Mujtaba. (n.d.). "Laporan Akhir Advokasi Media Korban Lumpur Lapindo." Yayasan Lapis Budaya Indonesia. unpublished.
Hamdi, Mujtaba. (2010). *29 Cerita Menentang Bungkam.* Surabaya: Penerbit Kanal.
Hilgartner, Stephen. (2007). "Overflow and Containment in the Aftermath of Disaster." *Social Studies of Science*, 37(1): 153–158.
Istadi, Bambang P., Gatot H. Pramono, Prihadi Sumintadireja, Syamsu Alam. (2009). "Modeling study of growth and potential geohazard for LUSI mud volcano: East Java, Indonesia." *Marine and Petroleum Geology* 26: 1724–1739.
Jajeli, Rois. (2015, May 28). "Kolam Penampungan Lumpur Lapindo Mencapai Luas 640 Hektar." *detikNews.*
Jasanoff, Sheila. (2007, June). "Bhopal's Trials of Knowledge and Ignorance." *Isis*, 98(2): 344–350.
JPNN.com. (2010, May 29). "Empat Tahun, Kasus Makin Kelam." http://www.jpnn.com/news/empat-tahun-kasus-makin-kelam (accessed August 22, 2017).
Keputusan Presiden Republik Indonesia Nomor 13 Tahun 2006. (2006).
Kabarsidoarjo. (2010, May 27). "Festival 4 Tahun Lumpur Lapindo." http://kabarsidoarjo.com/2010/05/27/festival-4-tahun-lumpur-lapindo/ (accessed August 25, 2017).
Kompas. (2011, May 29). "Warga Gelar Aksi 5 Tahun Lumpur Lapindo." http://regional.kompas.com/read/2011/05/29/11083550/Warga.Gelar.Aksi.5.Tahun.Lumpur.Lapindo (accessed August 25, 2017).
Kompas. (2015, May 30). "Tragedi Diperingati." *Kompas.*
Korbanlumpur. (2008, April 6). "Kisah Eef, Lowie Dan Anak Korban Lapindo," http://korbanlumpur.info/2008/04/kisah-eef-lowie-dan-anak-korban-lapindo/ (accessed August 25, 2017).
Knowles, Scott. (2014). "Learning from Disaster? The History of Technology and the Future of Disaster Research" *Technology and Culture*, 55(4): 7773–784.
Laksana, Yan Provinta. (2011). "5 tahun Lumpur Sidoarjo: Mencari Akar Masalah." http://www.kompasiana.com/yplaksana/5-tahun-lumpur-sidoarjo-mencari-akar-masalah_5500cdb1a333117f73511fcb (accessed August 25, 2017).
Lay, Cornelis. (2017, April). "Volunteers from the Periphery (Case Studies of Survivors of the Lapindo Mudflow and Stren Kali, Surabaya, Forced Eviction)." *Southeast Asian Studies*, 6(1): 31–61.
Lupi, M., E.H. Saenger, F. Fuchs, S.A. Miller. (2013). "Lusi mud eruption triggered by geometric focusing of seismic waves." *Nature Geoscience*, 6: 642–688.
Mardiyati, Ade. (2010, February 21). "Muddying the Truth: A New Documentary Looks at Sidoarjo." *Jakarta Globe.*
Mawuntyas, Dini. (2012, May 29). "6 Tahun Lumpur Lapindo, Siapa Sang Bethara Kala?" https://nasional.tempo.co/read/news/2012/05/29/058406883/6-tahun-lumpur-lapindo-siapa-sang-bethara-kala#ByvfgcqBwLGF8tTh.99 (accessed August 25, 2017).

Mazzini, A., H. Svensen, GG. Akhmanov, G. Aloisi, S. Planke, A. Malthe-Sorenssen, B. Istadi. (2007). "Triggering and dynamic evolution of the LUSI mud volcano, Indonesia." *Earth and Planetary Science Letters* 261: 375–388.

Mietzner, Marcus. (2014, October). "Indonesia's 2014 Elections How Jokowi Won and Democracy Survived." *Journal of Democracy*. 25(4), 111–125.

Muhtada, Dani. (2008). "Ethics, Economics and Environmental Complexity: The Mud Flow Disaster in East Java." *Systems Research and Behavioral Science*, 25: 181–191.

Nature. (2008). "Oil Company blamed for mud-volcano eruption." *Nature*, 456(6): 14.

NewScientist. (2008, November 8). "Did Drilling Cause Eruption?" *NewScientist*, 6.

Nuwer, Rachel. (2015, September 21). "9 Years of Muck, Mud and Debate in Java." *The New York Times*.

Novenanto, Anton. (2009). *Mediated Disaster The Role of Alternative and Mainstream Media in the East Java Mud Volcano Disaster.* (Unpublished Master's Thesis). University of Leiden, Leiden, The Netherlands.

Novenanto, Anton. (2010a, January–April). "Wacana Media Alternatif tentang Kasus Lapindo." *Jurnal Dinamika HAM* 10(1): 45–64.

Novenanto, Anton. (2010b, January–March). "Melihat Kasus Lapindo Sebagai Bencana Sosial." *Masyarakat, Kebudayaan dan Politik*. 23(1): 63–75.

Perrow, Charles. (1999). *Normal Accidents Living with High-Risk Technologies.* Princeton, NJ: Princeton University Press.

Putro, Herpin Dewanto. (2013). "Korban Lapindo: Suara Kami Takkan Pernah Padam…" http://nasional.kompas.com/read/2013/06/02/09190510/korban.lapindo.suara.kami.takkan.pernah.padam. (accessed August 25, 2017).

Setyarso, Budi, Heri Susanto, Zed Abidien, Rohmah Taufik. (2006, September 25). "A Bridge Over Muddy Waters." *Tempo English Edition*.

Schiller, Jim, Anton Lucas, Priyambudi Sulistiyanto. (2008). "Learning from the East Java Mudflow: Disaster Politics in Indonesia." *Indonesia*, 85: 51–77.

Reilly, Benjamin. (2009). *Disaster and Human History: Case Studies in Nature, Society and Catastrophe*. McFarland & Company.

Riski, Petrus. (2015, May 29). "Festival Budaya Tandai Peringatan 9 Tahun Tragedi Lapindo." https://www.voaindonesia.com/a/festival-budaya-tandai-peringatan-9-tahun-tragedi-lapindo/2797082.html (accessed August 25, 2017).

Riski, Petrus. (2016, June 1). "10 Tahun Lumpur Lapindo, Luka Itu Terasa Menyayat Hati." http://www.mongabay.co.id/2016/06/01/10-tahun-lumpur-lapindo-luka-itu-terasa-menyayat-hati/ (accessed August 25, 2017).

Samad, Herawanto. (2011). "5 Tahun Lumpur Lapindo." http://www.kompasiana.com/whaone/5-tahun-lumpurlapindo_551c186ca33311e22bb65a29 (accessed August 25, 2017).

Sawolo, Nurrochmat, Edi Sutriono, Bambang P. Istadi, Agung B. Darmoyo. (2009). "The LUSI mud volcano triggering controversy: Was it caused by drilling?" *Marine and Petroleum Geology* 26: 1766–1784.

Sawolo, Nurrochmat, Edi Sutriono, Bambang P. Istadi, Agung B. Darmoyo. (2010). "Was LUSI caused by drilling? –Authors reply to discussion." *Marine and Petroleum Geology* 27: 1658–1675.

Seyle, D. Conor, C. Siswa Widyatmoko, Roxane Cohen Silver. (2013). "Coping with natural disasters in Yogyakarta, Indonesia: A study of elementary school teachers." *School Psychology International*, 34(4): 387–404.

Subagyo. (2011). "Orang-orang "Berjasa" Bagi Bakrie: 5 Tahun Lumpur Lapindo." http://www.kompasiana.com/sbg/orang-orang-berjasa-bagi-bakrie-5-tahun-lumpur-lapindo_5500d1c2a33311237051216a (accessed August 25, 2017).

Suparno. (2017, May 24). "Korban Lumpur Lapindo Peringati 11 Tahun Semburan di Titik 21." *detikNews*.

Surya. (2014, May 26). "110 Patung Merana Ini Dipasang di Kolam Lumpur Lapindo." http://surabaya.tribunnews.com/2014/05/26/110-patung-merana-ini-dipasang-di-kolam-lumpur-lapindo (accessed August 25, 2017).

Steinberg, Ted. (2000). *Acts of God The Unnatural History of Natural Disasters in America*. Oxford, UK; New York, NY: Oxford University Press.

Tasic, Justyna, Sulfikar Amir. (2016). "Informational capital and disaster resilience: the case of Jalin Merapi." *Disaster Prevention and Management*, 25(3), 1–18.

Tejo, Amir. (2009, May 29). "Ribuan Korban Gelar Peringatan 3 Tahun Lapindo." https://news.okezone.com/read/2009/05/29/1/224159/ribuan-korban-gelar-peringatan-3-tahun-lapindo (accessed August 25, 2017).

Thorburn, Craig. (2009). "Livelihood Recovery In the Wake of The Tsunami Aceh." Bulletin of Indonesian Economic Studies. 45(1), 85–105.

Tingay, M.R.P., M.L. Rudolph, M. Manga, R.J. Davies, Chi-Yuen Wang. (2015). "Initiation of the Lusi mudflow disaster." *Nature Geoscience*, 8, 493–494.

Tingay, Mark, Oliver Heidbach, Richard Davies, Richard Swarbrick. (2008, August) "Triggering of the Lusi mud eruption: Earthquake versus drilling initiation." Geology 36 (8), 639–642.

Tribunnews. (2015, May 29). "Newsvideo: Peringati 9 Tahun Lumpur Lapindo Warga Arak Ogoh-ogoh." http://m.tribunnews.com/video/2015/05/29/newsvideo-peringati-9-tahun-lumpur-lapindo-warga-arak-ogoh-ogoh (accessed August 25, 2017).

Widyanto, Untung, Rohman Taufiq, Kukuh S. Wibowo. (2006a, June 19). "Mud Wrestling." *Tempo English Edition*, 40–41.

Widyanto, Untung, Sunudyantoro, Rohman Taifiq, Zed Abidien, Yosep Suprayogi. (2006b, August 28). "Negligence in Porong." *Tempo English Edition*.

Widyanto, Untung, Ahmad Fikri, Rohman Taufik, Zed Abidien. (2006c, October 23). "Porong Mud Fest." *Tempo English Edition*.

Zuber, H.M. Maksum. (2009). *Titanic Made by Lapindo*. Yogyakarta, Indonesia: Lafadl Pustaka.

PART III

Informational Relations

CHAPTER 7

Information Infrastructure and Resilience in American Disaster Plans

Megan Finn

Introduction

Resilience is a dominant paradigm for crisis planning in the context of the United States, and has been defined in White House Presidential Policy Directives as "...the ability to adapt to changing conditions and withstand and rapidly recover from disruption due to emergencies."[1] As part of this paradigm, plans for how information infrastructures will function in a crisis are steeped in the language of resilience. Information infrastructures are both important to the project of resilience as well as sites where the goal of resiliency is pursued. As the preceding definition of resilience suggests, in the US disaster-planning context, resiliency is predicated upon the ability to "rapidly recover" from emergencies. Yet, resilience is a complex concept with multiple meanings, and its efficacy as a dominant paradigm is not well understood, and even problematic for society's most vulnerable. This chapter takes historical examples of rapid recovery of information infrastructures after disasters, unpacks how resilience is imagined in contemporary plans for resilient information

M. Finn (✉)
University of Washington, Seattle, WA, USA
e-mail: megfinn@uw.edu

S. Amir (ed.), *The Sociotechnical Constitution of Resilience*,
https://doi.org/10.1007/978-981-10-8509-3_7

infrastructures, and asks what it means to recast repair activities as resilience. These examples illustrate the entanglements of the social and the material that are key to the concept of sociotechnical resilience (Amir and Kant 2018). Plans for resilient systems extend the project of securitization, but also must wrestle with capitalist logics. This tension reinforces that the concept of sociotechnical resilience must not only take into account the sociomaterial conditions of resilience, but also the complex institutional arrangements.

Disaster response policies emphasize a society's "resilience" as a way of identifying a community's ability to absorb change in the face of an implicitly troubling future. Resilience has positive connotations, but critics of resilience note that it ignores the power relations involved, which can produce uncertainty and an inequitable distribution of responsibility for sustaining resilience (MacKinnon and Derickson 2013; Tierney 2014, 2015; Walker and Cooper 2011; Weichselgartner and Kelman 2015; Welsh 2014). Kathleen Tierney (2014, p. 162) writes that there has been a multiplicity of definitions of resilience over the years, which has led to "a good deal of divergence in how the concept has been defined and operationalized by researchers." Furthermore, as noted by Tierney (2014, p. 162), the idea of resilience is not a particularly stable one—according to the definitions of resilience from the epigraph above, it is unclear what kind of recovery activities after a disaster would not be called resilient (Tierney 2014, p. 162). The polysemy of "resilience" is a subject of concern for many researchers in the field; its status as a boundary object both problematic and appealing (Tierney 2015). In light of this, I attempt to ground the term "resilience" in the policy documents that I discuss below, particularly using the definition cited above from the "White House Presidential Policy Directive (PPD-8)," which spawned many of the disaster response plans in place today. Yet, even in attempting to stabilize the definition of "resilience," questions of who should be resilient and on what time scale are difficult to ascertain from policy definitions of resilience, such as the one found in PPD-8.

Critiques of resilience are relevant to contemporary US disaster response policies attempting to enable resilience. In 2010, the United States federal government provided instructions for how local communities should engage in the disaster planning process.[2] In these planning guidance documents, resilience is explained as: "the ability to resist, absorb, recover from, or adapt to an adverse occurrence."[3] Based on these instructions, the work of creating resilience rests within a community and those in that community developing the disaster response plans: "Disasters begin and end locally. After the response is over, it is the local community that lives with the decisions made during the incident. Therefore, communities should have a say in how a disaster response occurs. They should also shoulder responsibility for building their community's resilience and enhancing its recovery before, during, and after a disaster."[4] In this formulation, the responsibility for resilience clearly lies in the work of a community.[5] Yet the power to create a resilient community may not lie within the community itself; sometimes the greatest disaster risks to a community come from outside sources, such as global warming. In the case of information infrastructures, outsiders also play a large role; the infrastructure operators may not be connected to the community and the community may have very little say over the resiliency of information infrastructures.[6] Those with the power to define and enhance the resilience of infrastructures and other sociotechnical assemblages are not generally the same people who suffer when an infrastructure does not "rapidly recover" after a disaster. Additionally, critiques of resilience explain that contemporary resilience policy is in fact a neoliberal construct, suggesting that rather than struggle against the system that causes the risk, resilience encourages people to accept what risks exist and take it upon themselves to brace for whatever comes next (Reid 2010; Tierney 2015). Resilience, according to these critics, is then a conservative approach that actually disempowers communities (MacKinnon and Derickson 2013).

Though scholars and policymakers have yet to settle on a specific definition of resilience, it has moved into a prominent role in crisis planning. However, its efficacy as an approach remains not well understood and potentially problematic. Tierney (2015, p. 3) summarizes the broad agreement that "Over approximately the past 15 years, the concept of disaster resilience has become increasingly central in research, policy, and practice." Volumes such as this one attempt to hone the more specific concept of sociotechnical resilience. Sociotechnical resilience requires that resilience is "understood as a double performance of social and technical systems" where the "social and technical are seamlessly integrated and mutually reinforced" such that a sociotechnical assemblage has the "capacity … for transforming itself from one configuration to another in the face of unforeseen crisis" (Amir 2018). Yet, critiques of the resilience concept must remain central in the formulation of sociotechnical resilience. The sociotechnical resilience of many infrastructures is not solely produced at the community level—the very nature of infrastructures is their reach beyond a single site of practice (Star and Ruhleder 1996). The neoliberal underpinnings of resilience prompt me to consider the conditions and capitalist relations under which infrastructures, and the uncertainty surrounding them, are maintained.

This chapter specifically considers the sociotechnical resilience of information infrastructures. As sociotechnical assemblages continually produced through the imbrication of the social and technical, information infrastructures (Bowker and Star 1999; Carse 2014) are critical for populations experiencing a disaster. The information practices associated with the continual production of public information infrastructures—networked resources that enable the circulation of information both constituted by and constituting publics—allow people to make sense of a disaster, as these practices are themselves sources for interpreting disaster; working and non-working infrastructure can answer questions about "how bad" a disaster is (Finn 2013). While the functioning of public information infrastructure is not a guarantor of physical well-being akin to anti-seismic building practices, the means to produce and circulate information is an important element of how people make sense of disasters and connect with others, contributing to their overall well-being. Furthermore, as Sulfikar Amir and Vivek Kant (2018) explain, "informational relations" are central to enabling sociotechnical resilience because they "constitute a key aspect of making humans and system hybrids," linking people with each other as well as technology. That is, even the project of creating sociotechnical resilience requires that the information infrastructure itself is resilient.

Over the next two sections, this chapter examines how repair of public information infrastructures in the face of crisis has worked in the past, and how the project of sociotechnical resilience is imagined in the United States' current infrastructure planning documents. In the final section, I then ask what it means to reframe infrastructural repair practices in terms of today's resilience policy.

Repair and Resilience

Over the past century and a half, public information infrastructures in Northern California have repeatedly had to repair and, to use the words of PPD-8, "rapidly recover" after earthquakes. Though the term "resilience" is a modern one used in contemporary disaster planning policy, in this chapter I highlight existing work that unintentionally falls under the broad definition of resilience, and what it means to pursue the objective of resilience.[7] This allows me to deliberately ask the question: *Given the broad definition of resilience found in Presidential Policy Directive 8, what does it mean to label infrastructure repair activities as acts of resilience?* Moreover, what does it mean to recast infrastructural repair as something to be planned for? And what does this analysis tell us about the project of sociotechnical resilience? In the following pages, I examine three historical examples of this resilience within the context of post-earthquake information infrastructure repair.

The 1868 Hayward Fault Earthquake

On October 21, 1868, at approximately eight o'clock in the morning, a major earthquake on the Hayward Fault near Oakland, California shook the Bay Area. The earthquake caused a great deal of damage in the small towns near the earthquake epicenter, as well as on "made ground" in San Francisco.[8] After the earthquake, the evening newspapers had to produce an edition even though their typesetting and printing rooms had been shaken by the temblor. The day after the earthquake, October 22, 1868, San Francisco's *Daily Evening Bulletin* wrote:

> Yesterday was one of the most difficult days to get the Bulletin out that we have ever experienced. The first shock of the earthquake paled all of the matter which had been set making it necessary to commence anew, after half-past 9 o'clock, to make up the paper. Then reports were industriously circulated to the effect that the Bulletin building was so much shattered by the shock that it was liable to fall at any moment. This rendered it almost impossible to induce compositors to work, and it was only after thorough examinations of the walls that men could be made to believe there was comparatively little danger. Then it was reported in the streets that the Bulletin would suspend its publications for the day. Notwithstanding this however, the Bulletin did appear, and kept its presses running until after 8 o'clock at night so great was the demand for information in regard to what had happened. Altogether, the Bulletin circulated 3,300 copies, and did not then supply the demand.[9]

The 1906 San Francisco Earthquake and Fire

Thirty-eight years later, on April 18, 1906, a large earthquake violently shook San Francisco again. Following the earthquake, fires burned for four days, destroying much of the city's business district. Many aspects of the city's telegraph and telephone systems, as well as the newspaper and post offices, were destroyed. Despite the fact that all the newspaper facilities in San Francisco had burned to the ground, a newspaper editorial declared that, "The maintenance of telegraphic communication with the outside world has been one of the most difficult matters with which the newspapers have had to contend."[10] Accounts of the efforts of telegraph operators and those who reestablished working telegraphic infrastructure appeared in the newspapers and telegraphy trade publications in the following days and months. A *Telegraph Age* report from May 1906 lionized the experiences of telegraph operators:

> The day following the earthquake, the Western Union Telegraph Company arranged to have a building erected at West Oakland, 100 by 50 feet in size, with an iron roof. In this building dynamos, gas engines, switchboards, quadrupled, duplex and Wheatstone machinery, a three-carload lot brought by Mr. L. McKisick, the electrician of the Western division, from Chicago, were installed and made an exact duplicate of the main office which had been destroyed in San Francisco. The current to operate the dynamos was secured from the Southern Pacific Railroad Company. Aerial cables were erected from the office to the city office at Oakland, where direct connection with all of the principal cities, and the wires in the cables leading over to San Francisco and through the underground system were used as legs to connect the branch offices in the ruined city direct with all important outside points. When it is considered that this office was not only established but built, an in full working operation within a week after the fire had destroyed the main office, it will be understood why the Western Union company resumed the prompt handling of all business to San Francisco and outside points.[11]

The 1989 Loma Prieta Earthquake

The most recent major earthquake in the San Francisco Bay Area, on October 17, 1989, famously occurred during a historic baseball World Series game featuring the San Francisco Giants and the Oakland A's. With its epicenter south of the Bay Area, the earthquake caused the most shaking and damage in Watsonville and Santa Cruz, and most of the fatalities occurred along a stretch of collapsed freeway in Oakland. Power failures and a lack of backup generators made it difficult for many radio, television, and newspaper companies to resume work immediately after the earthquake, but most were able to muddle through using generators or battery power.[12] Even though the information infrastructure of 1989 required a robust electricity grid, generators and battery power helped media companies continue to operate. In fact, many television and radio stations came through the disaster with enough working equipment to continue their broadcasts:

> When the earthquake hit, KCBS was 30 seconds away from the end of the national news feed and 2 1/2 minutes away from switching to live network coverage of the World Series. The station had dozens of reporters all around the San Francisco Bay region, an anchor booth at Candlestick Park, and another portable studio outside the park. The earthquake knocked out both the electrical power and most of the telephones at KCBS' offices in downtown San Francisco, but a rooftop emergency-power generator kicked in almost immediately, and the station came back on the air after only a few seconds. With the telephones to Candlestick Park out, the KCBS announcer could broadcast but did not know whether anyone was picking him up; and like many people at Candlestick Park, he did not know how severe the earthquake really was. ... About 6:15 p.m., the KCBS emergency-power generator failed, but engineers were able to tap into another generator, which gave the station a few lights in the broadcast studio and the ability to stay on the air. For the rest of the night, however, gallons of kerosene had to be hauled up 30 flights of stairs—the elevators were out—to keep the generator going.[13]

The efforts of KCBS's staff to keep their station on air illustrate just one example of the countless moments when, after the physical infrastructure broke, people fixed it so that it could continue to function. The employees at newspaper companies, telegraph networks, and radios stations repaired sociomaterial infrastructure: typesetters regained the strength (or were coerced) to return to work, electricians shipped parts from undamaged areas, and engineers fixed backup power to keep stations running. The repaired information infrastructure allowed people across the entire affected community to share their perceptions of the disaster, make their personal well-being known, and learn about how others had fared; moreover, it also enabled experts to circulate their explanations of what had happened and what to do next.

Across all of these historical moments, no one predicted the exact ways that the infrastructures would break in an earthquake, yet the work that was needed to repair them was rooted in past work practices and established understandings of what constituted "working infrastructure." In this sense, the repair of public information infrastructure was quite conservative—it reproduced what already existed (Finn in press). When disaster response planning had occurred, this also helped the repair of infrastructure; in 1989, the local government had purchased generators well in advance of the earthquake, presumably to ensure the continuity of infrastructure functioning. However, even with advance preparation, infrastructure repair always remain uneven; those with deep pockets can more easily throw resources into repair. And though these examples call attention to when information infrastructures have recovered quickly, there are also instances when infrastructure was not repaired.

Critical Infrastructure Planning

Today's federal disaster response plans in the United States imagines preserving public information infrastructures by making these infrastructures resilient. In our contemporary era of "preparedness," disaster planning is one technique amongst many that aims to ready society for a disaster (Lakoff 2007). US federal disaster plans specifically address this call for resilience in areas such as "public information" and "critical infrastructure." In this section, I examine how the 2013 *National Infrastructure Protection Plan* describes techniques for enhancing resilient qualities of the sociotechnical systems that underpin American life. I also look at one of the "critical" infrastructures highlighted in the plan—the information technology infrastructure. While the information technology critical infrastructure does not encompass all public information infrastructures in the United States, it does include things like the routing of internet traffic, a major underpinning of the contemporary public information infrastructure. I then discuss the current disaster plans for resilient infrastructures in light of the historic earthquake examples previously highlighted.

In 2011, President Barack Obama encoded the need for resilience into the "White House Presidential Policy Directive 8 (PPD-8)," which I quoted at the beginning of this chapter, noting that, "This directive is aimed at strengthening the security and resilience of the United States through systematic preparation for the threats that pose the greatest risk to the security of the Nation, including acts of terrorism, cyber-attacks, pandemics, and catastrophic natural disasters." PPD-8 resulted in the *National Preparedness System*, which includes five "frameworks" intended to help the United States prepare for different phases of a disaster: prevention, protection, mitigation, response, and recovery.[14] The disaster preparedness and mitigation framework documents contain the majority of the instructions about how communities can become resilient in the aftermath of a disaster. The Department of Homeland Security developed these plans and made them available to the public, including promotion through the Federal Emergency Management Agency's (FEMA) education courses.[15] The *National Infrastructure Protection Plan 2013: Partnering for Critical Infrastructure Security and Resilience* (*NIPP 2013*) document, the focus of the analysis below, is designed to work in tandem with the *National Preparedness System* set of frameworks.[16]

As outlined in *NIPP 2013*, infrastructures must become secure and resilient because they are foundational to the current way of life. *NIPP 2013* quotes the 2001 USA Patriot Act in defining critical infrastructure as representing "systems and assets, whether physical or virtual, so vital to the United States that the incapacity or destruction of such systems and assets would have a debilitating impact on security, national economic security, national public health or safety, or any combination of those matters."[17] Through the concept of resilience, plus the reconceptualization of information infrastructures as "critical," the plans outlined in *NIPP 2013* rely on government expertise to manage aspects of public information infrastructures denoted as critical, including defining what critical infrastructures are and techniques to identify to how to make the infrastructure resilient (Collier and Lakoff 2015). The invocation of these information infrastructures as sites of risk implicitly invites a particular planning logic into the management of these spaces (Adey and Anderson 2012; Collier and Lakoff 2015).

When information infrastructure is understood being at risk, then the problem becomes one of reducing risk. Critical infrastructure planning is mostly focused on improving resiliency through "technocratic solutions" (Tierney 2015, p. 9). As a part of risk mitigation, *NIPP 2013* describes a bureaucratic technology planning tool, the "Critical Infrastructure Risk Management Framework."[18] Yet, the agencies authoring these plans often don't have the ability to carry out such a risk management framework because many public information infrastructures are owned privately, an issue discussed throughout *NIPP 2013*.

While many federal disaster plans address emergency managers or people at various levels of government with some kind of responsibility for disaster response, *NIPP 2013* is directed towards "infrastructure owners and operators" as well as government entities and "other private and nonprofit organizations charged with securing and strengthening the homeland."[19] There are a number of ways in which *NIPP 2013* envisions collaboration with the private sector: "Government can succeed in encouraging industry to go beyond what is in their commercial interest and invest in the national interest through active engagement in partnership efforts."[20] The document contains many appeals to "a clear and shared interest in ensuring the security and reliance of the Nation's critical infrastructure."[21] There are also carrots such as, "the government can help private sector partners gain a more thorough understanding of the entire risk landscape" because "partners benefit from access to knowledge and capabilities that would otherwise be unavailable to them."[22] However, the incommensurable goals of the state and the private sector are readily acknowledged, as the state takes "non-economic values" into consideration, such that "government may have a lower tolerance for security risk than a commercial entity."[23] In the interest of drawing in private sector partners, the plan describes a number of organizational arrangements designed to foster information sharing across different infrastructure partners and the federal, state, local, tribal and territory governments.[24] *NIPP 2013* also warns private sector partners against sharing their infrastructure risk assessments publicly, since the Homeland Security Act of 2002 protects such assessments.[25] By securitizing the products of the risk assessments, this enables the state to assert some authority over privately owned infrastructures; in this case, the information products generated in the process of defining and analyzing infrastructure.

As infrastructure theorists make clear, all infrastructures are relational (Star and Ruhleder 1996). While *NIPP 2013* underscores that infrastructures are interdependent, the plan also separates different types of infrastructures in order to help simplify the problem of comprehending and managing potential risks. *NIPP 2013* named 16 different types of infrastructure, called out as Critical Infrastructure Sectors. Each Critical Infrastructure Sector has a corresponding Government Coordinating Council to organize all of the government agencies involved in running that particular infrastructure. *NIPP 2013* details the plans for managing the risks for each of these Critical Infrastructure Sectors, including establishing Sector Coordinating Councils, "self-organized, self-run, and self-governed" organizations that ideally "serve as a voice for the sector and represent principal entry points for the government to collaborate with the sector for critical infrastructure security and resilience activities."[26] Thus, the infrastructure sectors are defined by the state in *NIPP 2013*, but are operated at the will of the private infrastructure operators.

One Critical Infrastructure Sector constructed through *NIPP 2013* is for the IT Sector. The "Information Technology Sector-Specific Plan: An Annex to the National Infrastructure Protection Plan" represents a collaboration between private companies and government agencies through the Government Coordinating Council and the Sector Coordinating Council.[27] The voluntary nature of participation makes it a bit difficult to understand why particular partners are represented but not others. For example, major parts of the "cloud" are owned and operated by Amazon, who does not appear in the list of partners participating in the construction of the "Information Technology Sector-Specific Plan." This plan closely follows the Critical Infrastructure Risk Management Framework described in *NIPP 2013*.[28] This sector-specific plan describes a risk identification and assessment methodology, as well as "risk mitigation activity"—primarily technical solutions. The methods for assessing risk involve cataloging "threats" and apparent "vulnerabilities," as well as imagining "consequences." The last part of the plan is devoted to explaining how the IT sector might align their R&D priorities with the risk mitigation goals developed through the process of identifying "Risk Management Activities." That is, the risk assessment strategies, undertaken in the interest of resilience, are supposed to drive technology development. Thus, according to the "Information Technology Sector-Specific Plan," the ambiguously defined and understood project of creating resilience, and the associated techniques to enhance resilience, are shaping future technology project developments. However potent the discursive project seems to be, more research is needed to understand the implementation of these plans, how they influence activity, and if this actually contributes to resilience.[29]

Discussion

Given how promotion of infrastructural resilience is articulated in the United States' modern federal disaster plans, what does it mean for contemporary disaster response plans to rearticulate the kinds of sociotechnical repair activities described in these historical examples, within the paradigm of resilience? I argue that this rearticulation does three things:

First, using the work of Andrew Lakoff and Steven Collier, I show how "vital systems security" and resilience reconfigure information infrastructures as sites to be securitized. Securitization ideally, according to *NIPP 2013*, drives the development of information infrastructures. Second, following Lee Clarke (1999) and others, I highlight how disaster planning allows one group of people to symbolically exert power. Disaster planning attempts to put post-disaster infrastructure repair activities under the control of those who make resilience plans, usually the government agencies. Planning seeks to rationalize activities that might be very difficult to anticipate. In the resilience paradigm, the idea is not to exhaustively catalogue response activities, but to use risk management techniques to identify weaknesses and fix them. These modernist impulses are troubled by the problem of privately owned infrastructure. Third, I explore how the planning process operationalizes resilience by acknowledging that the responsibility for resilience lies with the (often private) owners of infrastructure—what the state calls their "partners"—who actually have authority over the infrastructures. Thus, resilience plans acknowledge the existence of risk and identify the methods for rationalizing it, but ultimately the state can only ask that communities and private organizations also adopt what disaster planning documents refer to as a "risk-conscious culture."[30] I read this as an example of the state attempting to control an aspect of privately owned information infrastructure in the name of protecting the homeland and its population, what Stephanie Kane, Eden Medina, and Daniel Michler (2015) call "infrastructural drift": "the behaviors and consciousness and unforeseen effects that accompany a systemic by systematic shift in technological habits." Infrastructural drift produces a challenge to security, which relies on a vision of the modern infrastructural ideal (Marvin and Graham 2001).

Infrastructure plans are interventions that "seek to increase the resilience of critical systems and to bolster preparedness for future emergencies" (Collier and Lakoff 2015, p. 22). Modern disaster planning recasts "critical" infrastructures as a site of risk to be managed. Critical infrastructures are an element of what Collier and Lakoff (2008a, b, 2015) call "vital systems security." Borrowing from Foucault's idea of biopolitics as a method of governing that aims to secure population well-being, and Beck's account of "reflexive risk" as modernity generating its own sources of destruction, Collier and Lakoff (2015) argue that vital systems security is a form of reflexive biopolitics. Contrary to Beck (1992), who argues that the complexity of risk is beyond traditional management techniques,

Collier and Lakoff argue that vital systems security attempts risk management precisely because of the complexity and interdependence of these systems and does not balk at the complexity of systems, but instead adapts governance approaches for these systems. In vital systems security thinking, it is not just disasters that are a source of vulnerability, but the very enabling-ness of "critical" infrastructures are problematic when we rely on them (Collier and Lakoff 2008b, 2015). Information infrastructure itself has been reconfigured as a site of risk because of both our dependence on it and the potential that it is broken.

The ideological origins of vital systems security in systems thinking are in part shared with that of resilience—resilience is symptomatic of complex systems thinking (Collier and Lakoff 2015; Welsh 2014). Resilience is then a solution to vital systems security because it promises to make systems that don't break, or at least recover quickly, like the newspaper, telegraph, or radio examples (Walker and Cooper 2011, p. 144; Welsh 2014, p. 18). Using vital systems security thinking, because information infrastructures are so important to resilience, and themselves must be resilient, these sociotechnical assemblages must be planned for and securitized. Building on the work of Collier and Lakoff, envisaging the examples of information infrastructure repair through the lens of securitization highlights that this ideology might not comport with the capitalist logics driving private sector infrastructure development.

Above I investigated how resilience is articulated through the *National Infrastructure Protection Plan*, though there are many other techniques to articulate and enact resilience.[31] Disaster planning more generally allows an organization to both assert authority over a domain that is unknowable and claim that they understand and can control it (Clarke 1999). This is obviously a position of great power. As Lee Clarke (1999, p. 41) explains, producing a disaster response plan gives a particular organization power over a problem set: "Symbolically, plans are public declarations that planners, or the planning organization, have thought carefully about some problem and have developed the necessary wisdom and power to establish dominion over it." And, of course, when one group of people has power, another necessarily does not: "Since claims to expertise are always claims that somebody should be left out of the decision loop, planning is deeply, unavoidably political." (1999, p. 13). Considering *NIPP 2013* from this perspective suggests that the planners—in this case the Department of Homeland Security—are asserting authority over infrastructure. Yet, this scenario is not so straightforward. *NIPP 2013* frequently refers to "partners," and shares the work of specific sectoral planning (as with the IT sector) with those partners. Still, the expertise to keep infrastructure running, which often resides in lowly engineers and technicians, is not easily accessed in the sweeping logics of these plans.[32]

An important function of a disaster response plan is that it can turn the unknowable into something manageable: "When organizations analyze problems, they try to transform *uncertainty* into *risks*, rationalizing problems previously outside the realm of systematic control" (Clarke 1999, p. 11). American disaster response plans simultaneously imagine and even prescribe actions, functioning as future-making but not future-determining documents. But disaster plans must be backwards-facing as well, looking to make the post-disaster future while preserving the functionality of the past. Being able to foresee all moments where quick infrastructural repair is necessary seems unlikely. The anticipatory techniques for achieving resilience—bundled together as risk management—involve cataloguing all of the ways that something could go wrong, or, in the words of *NIPP 2013*, determining the "threats" and "vulnerabilit[ies]." Disaster plans are thus an example of the "drive for rationalization in modern society" (Clarke 1999, p. 11). Disaster researchers have long been concerned with how disaster response planning over-exerts control without acknowledging the potential for creativity and improvisation that we see in instances of infrastructural repair (Harrald 2006; Wachtendorf and Kendra 2006; Weick 1998). In some sense, resilience is an effort to sidestep some of these issues. The idea of resilience is somewhat at odds with previous eras of preparedness, in that rather than catalog the possible scenarios and prescribe actions for acting in those scenarios, resilience asks that infrastructures achieve a particular state of constant "readiness" (Lentzos and Rose 2009). Resilience is predicated on the idea that a disaster cannot be predictable, and thus cannot be precisely planned for; the idea must be to achieve a more holistic approach. The idea that newspaper publishers must rally the typesetters to return to work, telegraph electricians have to go to herculean lengths to bring in supplies and repair equipment, or television engineers need to work around the clock to find yet another source of back-up energy and fuel supply, is ideally unnecessary in a resilient system because the possible sources of "risk" have been identified and mitigated.

Disaster response plans, as a form of "anticipatory action" to instate resilience, do not seek to prevent the hazards, but accept that they will happen and are a part of life (Anderson 2010). Yet, when we examine the techniques articulated in *NIPP 2013* for achieving resilience, they all involve rationalizing the present and attempting to anticipate the future (Anderson 2010). As the example of the sector-specific plan for information technology illustrates, managing infrastructures under the paradigm of resilience ideally sets the agenda for the development of technology going forward. But resilience is merely one of many values or goals that companies operating infrastructure might strive for.

Disaster plans reconfigure information infrastructure as a site of risk to be managed in the interest of increasing resilience. Though the state may exert the power to define the infrastructure, disaster plans alone do not give the state the power to induce resilience. Kane et al. (2015) illustrate this in their exploration of "infrastructural drift." Prior to the Chilean earthquake in 2010, called 27/2, "the military drifted away from maintaining its own radio broadcast (and telegraph) infrastructure and drifted toward privately built, for-profit digital communication infrastructures," and thus "the technohabits undermining military readiness" became "accompanied by only a shallow awareness or intent" (Kane et al. 2015). This kind of infrastructural drift can be extremely problematic when tsunami warnings do not arrive, as was the case in the 27/2 earthquake (Kane et al. 2015). The military didn't understand the communications infrastructure on which it relied, and when the time came, were not in a position to deliver crucial warning messages. Researchers have noted that the rise of "critical infrastructure protection" rhetoric coincided with "a period of intense reprivatization of formerly public infrastructural services, a move that later created an opportunity for secondary financial markets specializing in income streams (or securitized debts) arising from infrastructure privatization itself." (Walker and Cooper 2011, p. 153). The reprivatization of infrastructure, often called the "splintering" of infrastructure from the modern infrastructural ideal, makes it difficult for the state to fulfill its responsibility to citizens to ensure that infrastructures can "rapidly recover" after a disaster (Marvin and Graham 2001).

Applying the infrastructural drift concept in the United States could imply that there was once a moment where information infrastructures were publicly owned. Of course, the historical examples of infrastructure repair make it hard to idealize the "drift" from a publicly owned infrastructure to a privately owned one. Most of the examples of infrastructural repair mentioned previously feature employees working for privately held companies.[33] Yet just because an information infrastructure was privately owned, does not mean that it did not adopt the goals of the government. The Cold War is littered with examples of the mass media collaborating with the state on various forms of disaster planning (Knowles 2007). As Ben Anderson (2010, p. 780) states, the "anticipatory action is imbricated with the plurality of power relations that make up modern democracies." In light of the pluralized power and infrastructural drift, the IT sector plan written in conjunction with *NIPP 2013* represents an instance where the state wanted to collaborate with the private sector to further the state's goals of resilience. In analyzing this, I do not want to give over to ideology that says that the state should not interfere in the private business of the infrastructure owners. But I also don't want to assume that the state's interest in securitizing the homeland is not problematic. As Collier and Lakoff note, devoting state resources to vital systems security projects means not devoting resources and attention to other work.

When public information infrastructures are reimagined as sites of risk, then, according to the logic of modern disaster response plans, they also must then become resilient. The United States' federal disaster response plans encourage owners to make their infrastructures resilient through a set of activities that seek to identify and rationalize risk, then marshal the resources needed to address this risk. The work of addressing a risk might involve procuring personnel or tools for responding to a disaster, or making upgrades or repairs to existing infrastructures. The challenge for the project of sociotechnical resilience is that vast swaths of the public information infrastructure imbricated in the project of helping people make sense of a disaster are privately owned. While disaster response plans insert the state in instructing infrastructure owners about how to anticipate a disaster and communities about how to prepare for one, ultimately the disaster plans distribute the responsibility for responding to risk to private corporations, while also asserting the state's right to define what kind of infrastructure is "critical" and "at risk." This move, which attempts to make infrastructure resilient in peaceful times, allows the state to reassert itself in the work of defining what constitutes working infrastructure that is privately owned. Infrastructural drift poses a serious challenge to the work of vital systems security and sociotechnical resilience in federally led projects. Sociotechnical resilience must consider the complex institutional arrangements involved in producing infrastructure and other sociomaterial assemblages. For it is not simply the imbrication of the social and the material that is required to enable resilience, but also cooperation amongst multiple institutions with differing goals, expertise, and power.

Notes

1. White House. (March 30, 2011). "Presidential Policy Directive 8: National Preparedness," www.dhs.gov/presidential-policy-directive-8-national-preparedness

   The term resilience is described similarly in a later Presidential Policy Directive. "The term 'resilience' means the ability to prepare for and adapt to changing conditions and withstand and recover rapidly from disruptions. Resilience includes the ability to withstand and recover from deliberate attacks, accidents, or naturally occurring threats or incidents." White House. (February 13, 2013). "Presidential Policy Directive 21: Critical Infrastructure Security and Reliance," www.whitehouse.gov/the-press-office/2013/02/12/presidential-policy-directive-critical-infrastructure-security-and-resil
2. U.S. Department of Homeland Security. (2013). *National Infrastructure Protection Plan (NIPP) 2013: Partnering for Critical Infrastructure Security and Resilience*. Washington, DC: Department of Homeland Security, www.dhs.gov/publication/nipp-2013-partnering-critical-infrastructure-security-and-resilience; Federal Emergency Management Agency. (2010). *Developing and Maintaining Emergency Operations Plans: Comprehensive Preparedness Guide (CPG) 101*, Version 2.0.Washington, DC: U.S. Department of Homeland Security, www.fema.gov/media-library/assets/documents/25975
3. Federal Emergency Management Agency. (2011). *National Disaster Recovery Framework: Strengthening Disaster Recovery for the Nation*, www.fema.gov/media-library/assets/documents/24647; Federal Emergency Management Agency, *Developing and Maintaining Emergency Operations Plans*, 2–3.

   This issue of prescribing that others work to achieve resilience without any clear authority or enticement is pervasive throughout federal disaster response plans.
4. Federal Emergency Management Agency, *National Disaster Recovery Framework*; Federal Emergency Management Agency, *Developing and Maintaining Emergency Operations Plans*, 4–5.
5. Other disaster plans lay the responsibility for resilience in the hands of each group, or organization. For example, "Individuals, communities, NGOs, all levels of government, and the private sector should consider the long-term economic, health, social, and environmental dimensions of their choices and ensure resilience is maintained and improved. Sustainability employs a longer-term approach through plans, policies, and actions that reflect a comprehensive understanding of the economic, social, and envi-

ronmental systems within a community." U.S. Department of Homeland Security, *National Mitigation Framework*, 4.

One technique for ensuring a resilient community is the "Threat and Hazard Identification Guide (THIRA)." THIRA is supposed to cooperate well with the risk guidance in *NIPP 2013*. It is supposed to support the National Preparedness System by helping communities in "Identifying and Assessing Risks" as well as "Estimating Capability Requirements." THIRA outlines a four-step process whereby communities "identify threats and hazards of concern" in the context of the particular community and then the "capabilities" needed to address the threats. Then, the community is to identify shortcomings in the capability and draw on their resources and through partnerships with nearby communities, come up with ways to address the community's needs. Department of Homeland Security, *Threat and Hazard Identification Guide: Comprehensive Planning Guide (CPG) 201*, Second Edition, August 2013.

6. Collaborating with the private sector comes up in some other documents, particularly those about community planning: "Engaging the private sector is a critical element of the process. Much of the critical infrastructure necessary to our communities is owned and operated by the private sector. Connecting the government and the private sector is, therefore, a necessary part of the planning process." Federal Emergency Management Agency. (2010). *Developing and Maintaining Emergency Operations Plans*, 2–3.
7. There are many explanations of the genealogy of resilience. Most trace its origins to the ecological researcher C. S. Holling who wrote in the 1970s (Amir 2018; Walker and Cooper 2011; Welsh 2014).
8. Stover, C. W. and Coffman, J. L. (1993). *Seismicity of the United States, 1568–1989*, rev. ed., U.S. Geological Survey Professional Paper 1527. Washington, DC: United States Government Printing Office. Modified and abridged version retrieved from http://earthquake.usgs.gov/earthquakes/states/events/1868_10_21.php
9. "The 'Bulletin' Yesterday." (October 22, 1868). *San Francisco Evening Bulletin*.
10. "Newspapers Show Great Resources: Under Incredible Difficulties All but One of the San Francisco Dailies Survive." (April 29, 1906). *San Francisco Bulletin*.
11. "Telegraph Conditions in San Francisco." (May 16, 1906). *Telegraph Age* 24(10), 234, https://archive.org/details/TelegraphAgeV24
12. Subervi-Vélez, F. A. and Palerm, J-V. (1992). *Communicating with California's Spanish-Speaking Populations: Assessing the Role of the Spanish-Language Broadcast Media and Selected Agencies in Providing Emergency Services*. Berkeley, CA: University of California; Rapaport, R. (1998). "The Media: Radio, Television and Newspapers," in *The Loma Prieta, California*

Earthquake of October 17, 1989-Lifelines, edited by A. J. Schiff, United States Geological Survey Professional Paper 1552-A. Washington, DC: United States Government Printing Office, A43-46.

13. Rapaport, "Lifelines: The Media: Radio, Television and Newspapers."
14. U.S. Department of Homeland Security. (2011), *National Preparedness System*, www.fema.gov/media-library/assets/documents/29361; U.S. Department of Homeland Security. (2016). "National Planning Frameworks", www.fema.gov/national-planning-frameworks

    The National Preparedness System works with the National Incident Management System (NIMS), which sets forth guidelines for preparing for, responding to, and recovering from disasters. Resilience is defined consistently through these plans and frameworks. The *National Disaster Recovery Framework* plan similarly defines resilience as the "ability to adapt to changing conditions and withstand and rapidly recover from disruption due to emergencies." The definitions in these documents make it clear that the key ideas in resilience in disaster plans is adaptation to new circumstances and recovery. Federal Emergency Management Agency. (2011). *National Disaster Recovery Framework: Strengthening Disaster Recovery for the Nation*, 81; Federal Emergency Management Agency. (2010). *Developing and Maintaining Emergency Operations Plans.*
15. Federal Emergency Management Agency. (2015). "IS-860.c: The National Infrastructure Protection Plan, An Introduction," www.training.fema.gov/is/courseoverview.aspx?code=IS-860.c
16. US Department of Homeland Security, *National Infrastructure Protection Plan (NIPP) 2013*. This plan supersedes 2009 and 2006 plans. *NIPP 2013* has the following resilience-oriented "vision:" "A Nation in which physical and cyber critical infrastructure remain secure and resilient, with vulnerabilities reduced, consequences minimized, threats identified and disrupted, and response and recovery hastened." Ibid., p. 5. All of the concepts mentioned here—security, resilience, vulnerability, and threats—have distinct histories and are deservedly the subject of books and journal articles.
17. Ibid., p. 7.
18. Ibid., p. 15. This framework asks its users to attend to both the goals and objectives set in NIPP 2013 and future "calls to action," which include "joint priority setting." Next, the users of the framework are instructed to identify all of the aspects of their infrastructure that are important to functioning, including interdependencies with other infrastructures. In the process of examining their identified infrastructure, users are supposed to identify the (1) threats, such as earthquakes or terrorism, (2) vulnerabilities, which include "physical feature[s] or operational attribute[s] that render[] an entity open to exploitation or susceptible to a given hazard," and (3) consequences of an event. The "Critical Infrastructure Risk

Management Framework" then describes how infrastructure operators should "implement risk management activities," as an approach to "mitigate consequences." Activities such as "work to restore critical infrastructure operations following an incident," and "repair or replace damaged infrastructure with cost-effective designs that are more secure and resilient," are strategies for resilience. These general descriptions in the plan might correspond to post-disaster infrastructural repair activities such as the fixes to the newspaper, telegraph and radio described above. The last step of the framework is to "measure effectiveness."

19. Ibid.
20. Ibid., p. 2. This theme is discussed many times throughout the document: "The NIPP 2013 acknowledges that the Nation's critical infrastructure is largely owned and operated by the private sector; however, Federal and SLTT governments also own and operate critical infrastructure, as do foreign entities and companies."
21. Ibid., p. 1.
22. Ibid., p. 10.
23. Ibid., p. 10.
24. Ibid., Appendix A.
25. "The Federal Government has a statutory responsibility to safeguard critical infrastructure information. [footnote: Under the Homeland Security Act of 2002, §201(d)(11)(a), DHS must ensure that any material received pursuant to this Act is 'protected from unauthorized disclosure and handled and used only for the performance of official duties.'] DHS and other agencies use the Protected Critical Infrastructure Information (PCII) program and other protocols such as Classified National Security Information, Law Enforcement Sensitive Information, and Federal Security Classification Guidelines. The PCII program, authorized by the Critical Infrastructure Information (CII) Act of 2002 and its implementing regulations (Title 6 of the Code of Federal Regulations Part 29), defines both the requirements for submitting CII and those that government agencies must follow for accessing and safeguarding CII." U.S. Department of Homeland Security, *National Infrastructure Protection Plan*, 17–18.
26. Ibid., p. 35. Not all of the sectors seem as publicly organized as the IT sector. For example, the communications sector did not have an updated plan publicly available as of March 2016.
27. The state is represented by the Government Coordinating Council (GCC) and the private sector by the Sector Coordinating Council (SCC).
28. The "Information Technology Sector-Specific Plan" adopts an approach to risk that starts with identifying the "critical functions" of the IT infrastructure. These critical functions include: the provisioning of IT products and services, domain name resolution and routing of Internet traffic, "incident

management capabilities," "identity management and trust support services," and making available Internet-based content, information and communication. Risk is identified relative to this functionality.

29. The project of Cold War–era Civil Defense is possibly instructive. While the project of Civil Defense focused on preparing Americans for nuclear war is largely understood to be a failure, the program succeeded at setting up the long reach of federal disaster management into the states. This suggests that the value of the critical infrastructure project might not necessarily be simply measured in changes to infrastructure in the interest of resilience, but the implications of activities such as gathering private and public sector "partners" together (Knowles 2007, 2011; Rozario 2007).
30. U.S. Department of Homeland Security, *National Mitigation Framework*.
31. Other work examines how this concept is articulated in other types of activities such as simulation and table-top exercises, (see, for example, Adey and Anderson 2012; Lakoff 2007).
32. According the Clarke, even as planning creates experts that excludes others, planning, and more importantly, the claims over a domain that come with planning, crowd out the possibilities of other kinds of action. A host of other stakeholders are left out of the planning process—those who make use of the infrastructures, for example.
33. Yet, much of the development of infrastructure mentioned in the IT sector-specific plan was originally funded by the state under the National Science Foundation Network. In the 1980s and 1990s, control of the Internet backbone was turned over to private companies, and much of the media industry was deregulated by the Telecommunications Act of 1996 (McChesney 2013).

References

Adey, P., & Anderson, B. (2012). Anticipating emergencies: Technologies of preparedness and the matter of security. *Security Dialogue*, *43*(2), 99–117. https://doi.org/10.1177/0967010612438432

Amir, S. (2018). Introduction: Resilience as Sociotechnical Construct. In S. Amir (Ed.), *The Sociotechnical Constitution of Resilience: A New Perspective on Governing Risk and Disaster*. Palgrave Macmillan.

Amir, S., & Kant, V. (2018). Sociotechnical Resilience: A Preliminary Concept. *Risk Analysis*, 38: 8–16. https://doi.org/10.1111/risa.12816

Anderson, B. (2010). Preemption, precaution, preparedness: Anticipatory action and future geographies. *Progress in Human Geography*, *34*(6), 777–798. https://doi.org/10.1177/0309132510362600

Beck, U. (1992). *Risk society: towards a new modernity*. London: SAGE Publications.

Bowker, G. C., & Star, S. L. (1999). *Sorting things out: Classification and its consequences.* Cambridge, MA: MIT Press.
Carse, A. (2014). *Beyond the big ditch: Politics, ecology, and infrastructure at the Panama Canal.* Cambridge, MA: MIT Press.
Clarke, L. (1999). *Mission improbable: Using fantasy documents to tame disaster.* University of Chicago Press.
Collier, S. J., & Lakoff, A. (2008a). Distributed preparedness: The spatial logic of domestic security in the United States. *Environment and Planning D: Society and Space, 26*(1), 7–28. https://doi.org/10.1068/d446t
Collier, S. J., & Lakoff, A. (2008b). The vulnerability of vital systems: How "critical infrastructure" became a security problem. In M. D. Cavelty & K. S. Kristensen (Eds.), *Securing "The Homeland": Critical infrastructure, risk and (in)security.* New York: Routledge.
Collier, S. J., & Lakoff, A. (2015). Vital systems security: Reflexive biopolitics and the government of emergency. *Theory, Culture & Society, 32*(2), 19–51. https://doi.org/10.1177/0263276413510050
Finn, M. (in press). *Documenting aftermath.* Cambridge, MA: MIT Press.
Finn, M. (2013). Information infrastructure and descriptions of the 1857 Fort Tejon Earthquake. *Information & Culture: A Journal of History, 48*(2), 194–221. https://doi.org/10.1353/lac.2013.0011
Harrald, J. R. (2006). Agility and discipline: Critical success factors for disaster response. *The Annals of the American Academy of Political and Social Science, 604*, 256–272.
Kane, S. C., Medina, E., & Michler, D. M. (2015). Infrastructural drift in seismic cities: Chile, Pacific Rim, 27 February 2010. *Social Text, 33*(1 122), 71–92. https://doi.org/10.1215/01642472-2831880
Knowles, S. G. (2007). Defending Philadelphia: A historical case study of civil defense in the early Cold War. *Public Works Management & Policy, 11*(3), 217–232. https://doi.org/10.1177/1087724X06297346
Knowles, S. G. (2011). *The disaster experts: Mastering risk in modern America.* Philadelphia: University of Pennsylvania Press.
Lakoff, A. (2007). Preparing for the next emergency. *Public Culture, 19*(2), 247–271. https://doi.org/10.1215/08992363-2006-035
Lentzos, F., & Rose, N. (2009). Governing insecurity: Contingency planning, protection, resilience. *Economy and Society, 38*(2), 230–254. https://doi.org/10.1080/03085140902786611
MacKinnon, D., & Derickson, K. D. (2013). From resilience to resourcefulness: A critique of resilience policy and activism. *Progress in Human Geography, 37*(2), 253–270. https://doi.org/10.1177/0309132512454775
Marvin, S., & Graham, S. (2001). *Splintering urbanism: Networked infrastructures, technological mobilities and the urban condition.* London: Routledge.

McChesney, R. W. (2013). *Digital disconnect: How capitalism is turning the Internet against democracy*. New York: The New Press.

Reid, J. (2010). The disastrous and politically debased subject of resilience. Presented at the Symposium on the Biopolitics of Development: Life, Welfare, and Unruly Populations, Calcutta.

Rozario, K. (2007). *The culture of calamity: Disaster and the making of modern America*. Chicago: University of Chicago Press.

Star, S. L., & Ruhleder, K. (1996). Steps toward an ecology of infrastructure: Design and access for large information spaces. *Information Systems Research*, *7*(1).

Tierney, K. (2014). *The social roots of risk: Producing disasters, promoting resilience*. Stanford University Press.

Tierney, K. (2015). Resilience and the neoliberal project: Discourses, critiques, practices—and Katrina. *American Behavioral Scientist*, *59*(10), 1327–1342. https://doi.org/10.1177/0002764215591187

Wachtendorf, T., & Kendra, J. M. (2006). Improvising disaster in the City of Jazz: Organizational response to Hurricane Katrina. In *Understanding Katrina: Perspectives from the Social Sciences*. Social Science Research Council. Retrieved from http://understandingkatrina.ssrc.org/Wachtendorf_Kendra/

Walker, J., & Cooper, M. (2011). Genealogies of resilience: From systems ecology to the political economy of crisis adaptation. *Security Dialogue*, *42*(2), 143–160. https://doi.org/10.1177/0967010611399616

Weichselgartner, J., & Kelman, I. (2015). Geographies of resilience: Challenges and opportunities of a descriptive concept. *Progress in Human Geography*, *39*(3), 249–267. https://doi.org/10.1177/0309132513518834

Weick, K. E. (1998). Introductory essay: Improvisation as a mindset for organizational analysis. *Organization Science*, *9*(5), 543–555.

Welsh, M. (2014). Resilience and responsibility: governing uncertainty in a complex world: Resilience and responsibility. *The Geographical Journal*, *180*(1), 15–26. https://doi.org/10.1111/geoj.12012

CHAPTER 8

An Audience Perspective on Disaster Response

Kurniawan Adi Saputro

Introduction

It has been established in the literature of disaster sociology (Fritz and Williams 1957; Quarantelli 1994, 2003) that the first responders, those who come first to rescue, evacuate, and help those affected by disaster, are the people themselves. They can be neighbours, families, and even strangers to the victims and survivors. This phenomenon has been observed across times and geopolitical boundaries that some claimed it to be a "natural" response to a disaster (e.g., Quarantelli 2008). It is characteristically pro-social, in contrast to the response toward conflicts. In this light, we need to ponder why some disaster events received extraordinary responses, while some others received lukewarm reactions. For example, in 2010 there was a rare incident of two big disasters occurring in the same period in Indonesia. On October 25, 2010 a 7.7 Richter scale earthquake struck the Mentawai Islands (Cedillos and Alexander 2010) and cost 456 lives. A day later, Mount Merapi started a series of eruptions that took 332 lives.

K. A. Saputro (✉)
Indonesia Institute of the Arts Yogyakarta, Yogyakarta, Indonesia
e-mail: kurniawan.as@isi.ac.id

S. Amir (ed.), *The Sociotechnical Constitution of Resilience*,
https://doi.org/10.1007/978-981-10-8509-3_8

If we examine the media coverage of the magnitude of an event by counting the lives lost to it (Knight 2006), the Mentawai Island disaster is larger and, by implication, worth more media exposure than the Mount Merapi disaster. It turned out that the Indonesian media in general quickly shifted and focused their attention from Mentawai to Merapi. The reason is simple: Merapi is located at the heart of Java Island, where most of the Indonesian population is concentrated, whereas the Mentawai Islands are in the periphery off Sumatra Island. As a result, along with the shift of media attention, the surge of help and aid went to Mount Merapi.

Although the apparent discrepancy of attention between these two events is interesting in itself, this chapter is not going to address that issue. Instead, the case is illustrative of an inextricable link between disaster events and media coverage, and its subsequent audience response. To the majority of Indonesian media audiences, those disaster events were made real to them through media. Even to those living in the areas surrounding Mount Merapi, the physical phenomena were experienced, partly, through amateur and commercial radio reports, Twitter updates, print media articles, and news coverage on national television channels. This media coverage provides a necessary condition for citizen's collective attention to the disaster event. This media coverage is necessary but not sufficient for audiences' collective action to help those affected. This chapter is going to outline the current trend in media developments in Indonesia in particular, and the world in general, that affect how citizens decide to join in disaster relief efforts. It then argues that a closer look at disaster communication from the audience perspective can shed light on the processes audiences experience, and the obstacles involved, that start from their collective attention and end with collective action. Understanding disaster relief response through audience perspective becomes mandatory now because the world is more and more deeply connected through media that, in general, our reactions toward mediated disaster are preconditioned by our place in the media environment. Our capacity to act collectively, partially, is also related to the kind of media we can use as a collectivity.

Contemporary Audiences of Mediated Disaster

In this section I want to connect the point I made earlier about people's critical role as first responders in disaster response with the nature of contemporary media. I want to make use of observations that have been made by different authors (Webster 2005; boyd 2011) regarding the ubiquity of media and publicness enabled by certain affordances of the new media. In so doing, I want to argue that although people's inclination to help in response to disaster is universal, current engagement with media allows for larger organization and, perhaps, better performance of such helping behaviour. In other words, current media may function to amplify audiences' response to disaster.

By the second half of the twentieth century, it has been commonplace to see rapid growth of media outlets and their expansion to new platforms. Although there is a trend of consolidation of media into bigger corporations, new media companies are still built and capture economically feasible market share. Personal computers and mobile devices become available in ever-expanding areas and their price steadily becomes cheaper. These trends only add to the ubiquity of media in our everyday lives. The other side of ubiquity is fragmentation, in which no one media outlet can command audiences' collective attention. Where in the past a national audience could be viewed as uniformly exposed to a limited number of national television channels, now major channels in the United States only have around a 10 per cent share of the national audience (Webster 2005). My own study of Indonesian audiences during the Mount Merapi eruption in 2010 (Saputro 2014) shows that their collective attention was fairly distributed (fragmented) into different media outlets during the two months of crisis. This means that audiences' collective attention moved rather easily from one media outlet to another. If these diverse media outlets somehow decided to promote their own agenda for their own audience, there would not be a collective attention toward the Mount Merapi eruption and its impact. Instead, various media that served different segments on various platforms decided to focus on the Mount Merapi eruptions, although only for a relatively short period of time and 'excluding' the aftermaths of the Mentawai Islands tsunami. Fragmentation of media outlets does not always lead to fragmentation of issues.

The second observation made by boyd (2011) is that the structural properties of social network sites allow for a new form of sociality, a public sociality. Three of the properties are persistence, scalability, and searchability (p. 46). Persistence refers to the new platform's ability to store and archive information, scalability means that media users have bigger opportunities of being visible to the world, and searchability refers to the platform's ability to search all the information ever recorded in its system. These properties are now common in almost all social network sites, whether they are video-based (e.g., YouTube), image-based (e.g., Instagram), text and image-based (e.g., Twitter), or multimodal message (e.g., Facebook). In the context of disaster response, these three capabilities mean information that might be critical to those affected by disaster can be entered into the system by many people simultaneously, stored, viewed, and searched by many people in different locations at the same time. These media platforms have been used to distribute crisis-related messages (Bruns and Burgess 2012; Bruns et al. 2012), to connect different sources of information (Starbird and Stamberger 2010; Starbird 2011), to map needs on the ground (Meier and Munro 2010), and to connect the affected communities and relief donors (Saputro 2016). In a comprehensive review of literature (academic and popular) on the use of social media during disaster, Houston et al. (2014) found fifteen types of use before, during, and after the disaster event.

It was the adoption of social media in particular and the Internet on smartphones in general that enabled the emergence of Jalin Merapi as a popular information source and hub of disaster relief efforts for the people. As a comparison, an earthquake on the southern side of Yogyakarta four years earlier received a similar outpouring of responses, but it did not show a similar level of organization although both received very intense mainstream media coverage. Yet, the crucial role of social media in disaster, I would argue, should not lead us to assume that social media were used exclusively by audiences or that social media were the most effective tool for disaster relief effort (see also Carah and Louw 2012; Ewart and Dekker 2013). On the contrary, we will reap better and more realistic insights when analyzing different media that audiences use in responding to a disaster event. In my study (Saputro 2014), I examined media outlets audiences used during the Mount Merapi eruptions and found that national television channels were the most popular media outlets to all age groups. Traditional mass media were very important for all audiences. However, the younger demographic used Facebook, Twitter, and frequented the Jalin Merapi website, whereas the older demographic did not visit social network sites. The older demographic read print newspaper, whereas the younger demographic did not. These similarities and differences between different age groups provide us insights into how they took action toward the disaster. Both groups donated to the refugees, but the older group channeled their donations through media they trusted, whereas the younger group preferred sending donations directly to the refugees or through some connections they preferred. Of course it was not a clear-cut difference, and there were overlaps. Further, young people, the users of social media, are the dominant group of the volunteers.

To see disaster response from the perspective of audience, we need to take into account several issues that pertain to how the audience receives and makes meaning out of disaster-related media messages. I would like to propose a framework that sees audiences' response to mediated disaster as a move from collective attention to collective action. Here their collective nature does not merely result from being aggregated via discursive technological means such as rating surveys, but, more importantly, it results from having real or imaginary relationships with each other and having a common purpose toward those affected by disaster. Hence, they become a collective entity through taking intentional acts toward those affected by disaster, rather than through exposure to messages on media (Webster et al. 2006).

From Collective Attention to Collective Action

The framework I am developing anticipates several issues that have been identified by different scholars, such as fragmentation of media use (Webster 1986, 2002, 2005), action at a distance (Boltanski 2004), audiences' relationship with the suffering others (Chouliaraki 2006), and mediated orientation to public issues (Couldry et al. 2007). These issues are relevant to the audiences' collective experience because their mediated relationship with the disaster created specific sets of problems, which will be clarified in the following paragraphs.

First, as I have mentioned in the opening paragraphs, proliferation of media devices and media outlets has fragmented the use of media into separate clusters of use based on habits and/or preferences. When audiences spread their attention to many media outlets through different media devices, a common characteristic in developing and developed countries, they pay attention to different issues chosen by different media, making collective action harder to materialize. Different choice of issues, segments of audience markets, and cycles of information production can prevent audiences from having an overlap of interests.

Second, audiences are physically separated from each other and from the people affected by disaster. They cannot immediately act on the problem depicted in the media, do not know who among them are interested in taking an action or how to do it. Thus, the physical separation amongst audiences and between audiences and the mediated others needs to be overcome. Each sociotechnological solution that is used to overcome the problem entails its own specific problems. For example, the relay of money from audiences to the refugees via a newspaper organization creates its own problems of accountability and confusion of interests, among others.

Third, generally speaking, the audiences and the survivors of disaster are strangers to each other. Here we are not talking about the relationship between an individual audience member and an individual survivor, but between both collectivities. Mainstream mass media are not designed to maintain personal connection among their audiences on a regular basis. On the contrary, social media are used exactly to connect with each other. However, these connections revolve around persons and personalities (boyd 2011), rather than public issues (Adams 2012). Further, social media are generally used to maintain connection between families and friends, rather than to forge connections among strangers (Adams 2012).

Fourth, orientation to public issues through the use of media is not necessarily followed by public action. Links and opportunities to connect the starting point, namely engaging public issues through media, with effective deliberation or action do not sufficiently exist. These can be values that bridge private-public worlds or the right social context for talking about the issues, as identified by Couldry et al. (2007), or an extraordinary event such as a disaster that provides the impetus to act.

The model I am developing results from approaching the trajectories from attention to action as a series of barriers that confront media audiences. In other words, they become barriers only for a specific social behavior, namely the use of media. For example, attention fragmentation is only an obstacle when we consider how people are collectively exposed to increasingly abundant media outlets. These issues do not exist in a period or in a society where media outlets are short in supply. In this perspective, the barriers that confront media users are:

1. audiences' attention do not naturally focus on a disaster;
2. salient features of the disaster are differently formulated;
3. audiences' mental picture of the affected people and their relationship to them are not obvious;
4. resources needed to take action are not necessarily available; and
5. boundaries between private and public domain present specific challenges to a collective action.

By viewing the transformation of attention to action as efforts to overcome these barriers, I can conceptualize the processes in the diagram

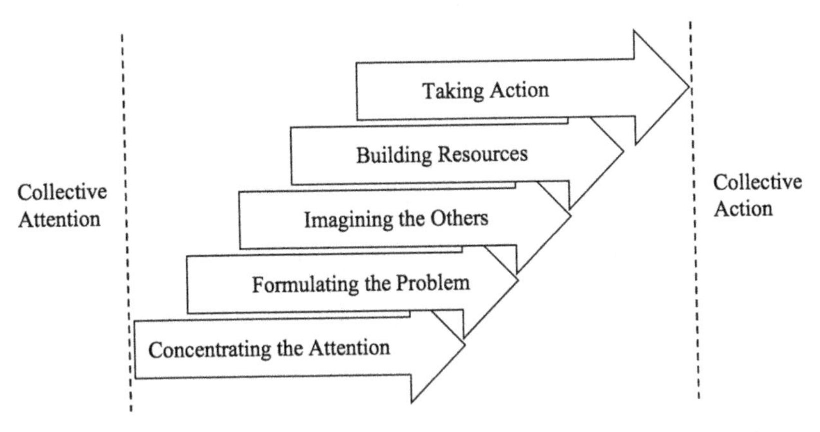

Fig. 8.1 Processes of transformation from collective attention to collective action

below (Fig. 8.1). The diagram shows collective attention and collective action as being separate. The processes to move from attention to action, or to overcome the barriers, are represented using arrows. Rather than arranging the arrows sequentially to show a linear progression, I stack them on top of each other to show that each process builds on top of the other processes although each process progresses independently. It also indicates that one process does not end when the other process starts. Each process will be described and discussed in a separate section, starting from the bottom and moving upward. To clarify the points, each process will be illustrated with my study of audiences' response to the Mount Merapi eruption in 2010 (Saputro 2014). The following paragraphs will describe a short chronology of the disaster events, and then elaborate on different processes involved in the transformation of collective attention to collective action.

After being declared a Level III Emergency five days before and a Level IV (the highest) one day before, Mount Merapi erupted on October 26, 2010. Surono, the head of the Indonesian Center of Volcanology and Geological Hazard Mitigation, took over the monitoring from a local observatory center (BPPTK) and served as the key government authority during the crisis (Kurnia 2014). Although the warning had been issued timely, the first explosion killed thirty-five people who refused to leave, including Maridjan, the mountain's mystical guardian and a quite well-known media star for refusing the Yogyakarta's governor order to evacuate and surviving the previous eruption in 2006. In these first days, local and national media paid heavy attention to whether or not Maridjan survived. Subsequently, the media focused on the uncertain livelihood of refugees, who numbered more than three hundred thousand people, as the danger zone was expanded from ten to fifteen, then to twenty kilometers. Finally, the peak explosion occurred on November 5, 2010, which took the lives of 332 people despite timely evacuation.

Amidst the chaotic circumstances, the public found that there was a grassroots information network called Jalin Merapi that provided frequent updates of the volcanic crisis and the refugees. Jalin Merapi was founded in 2006 by three community radio stations on the slopes of Mount Merapi, two community radio networks, and four local NGOs in the neighbouring cities to provide the people of Merapi with information appropriate to their local needs, such as early warning systems and disaster mitigation measures (Tanesia and Habibi 2011). This network swiftly moved to recruit volunteers and to station them close to the refugees. Their main duty initially was to survey and report the refugees' needs, but later they distributed aid for refugees as well. Furthermore, the community radios of Jalin Merapi intensified their regular monitoring of Mount Merapi's crater and reported its visual characteristics to their new audiences, since their handheld transceiver communication was now streamed to the world through the Internet by a Yogyakartan NGO specializing in information and communication technology. The intense report from national and local media resulted in the massive influx of relief aid from individuals and organizations, which was, to an extent, pooled and distributed by the media themselves. To give a sense of the scale, on the peak of the eruption a local newspaper recorded 180 donations on its daily list. The media attention also brought an unintended consequence when a national gossip show on television spread an unsubstantiated rumor of a larger, impending explosion only two days after the last, biggest explosion in Mount Merapi's centennial history (Surono et al. 2012). The rumor spawned a quick and severe backlash from television audiences in the form of complaints to the Indonesia Broadcasting Commission. These three types of mediated collective action—volunteering, donating, and complaining—became key cases upon which I build my conceptual framework, which I will detail later in the chapter. The aim, again, was to understand how people collectively respond to a disaster in the age of saturated media.

Concentrating the Attention

From audiences' point of view, the proliferation of media outlets entails the problem of abundance. On an individual level, the problem of abundance simply means that an individual's time to consume media cannot match with media offerings. In aggregate, the rising supply of media content and market differentiation lead to audience segmentation into different socioeconomic groups each served with different issues, contents, and perspectives. On the one hand, media compete with each other for audiences' attention. On the other hand, the audience themselves cannot expect that they pay attention to the same issue.

My study of 539 Indonesian media audiences during the Mount Merapi crisis found that their collective attention to various media outlets across platforms was relatively fragmented (Saputro 2014). To further complicate the matter, there was another disaster in the same period as Mount Merapi's eruption, which could be another distraction to audiences' collective attention. Moreover, the National Disaster Management Agency (BNPB) declared that the status of the Mount Merapi disaster was local, which meant that the local government had the capacity to handle the impacts. In other words, national media did not have to focus on the Mount Merapi disaster continuously because it was a "local" event. These reasons could have resulted in the media paying attention to diverse issues. It turned out that Mount Merapi became the central issue in the national media for at least several weeks.

One possible explanation was that audience of Indonesian media is concentrated in Java (Nugroho et al. 2013). Plausible as it is, we still need to understand whether audiences are actually receptive of media coverage. Without audiences' reception, media's issue will not move them into collective action. We can get a glimpse of audiences' reception of media coverage from their response. I collected data about audiences who took action by volunteering to Jalin Merapi, giving donation for refugees through a local newspaper, and filing complaints to the Indonesia Broadcasting Commission about a television show that spread rumors about an impending catastrophe during the peak of the crisis. Figure 8.2 shows the number of audiences who took action along the axis of time. October 26, 2010 is the start of the first series of eruptions and November 5, 2010 is the start of the second, much bigger, series of eruptions. Applications to Jalin Merapi (blue line) closely parallel the progression of the eruption, with spikes of application follow the peak of eruption. Donation via a local

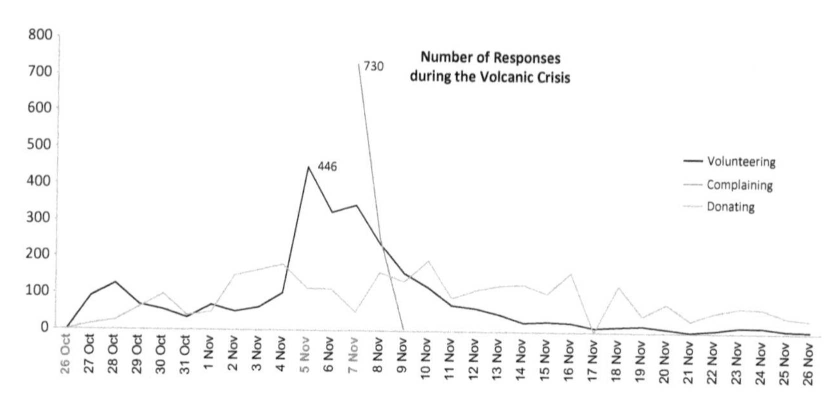

Fig. 8.2 Dynamics of three forms of collective actions toward mediated disaster

newspaper (green line) shows a rising trend although it is not sensitive to the dynamics of the event. The complaints filed to the Indonesia Broadcasting Commission peaks on the same day the infamous show airs and declines rapidly.

Based on my survey and interviews, more than half of those who volunteered and those made complaints knew about the opportunity to act directly from the media. They followed the progression of events through media, whether it was citizen-led media such as Jalin Merapi or any other traditional media (Nielsen 2010). Despite the fact that their collective exposure to media was fragmented, the media themselves focused on the same issue. On the other hand, audiences chose to follow that issue on different media outlets. To put if differently, media's focus on an issue does not necessarily lead to audiences' focus on that issue. But when the two converged, they created a "sharedness" (Dayan 2009). The feeling of sharing the same world was facilitated by how media manage temporal aspects of their reporting, such as using schedules, loops, and live coverage.

Formulating the Problem

The second problem faced by audiences is how the event is characterized as a problem. It is a critical stage because a disaster event, especially to the audiences who live too far from it to be affected directly, can be seen as many things or even "nothing." For example, Chouliaraki (2006) typified news about suffering as "adventure," "emergency," and "ecstatic news," each offering different possibilities of identification and agencies. Fine and White (2002) identified four processes that enabled a human interest story to establish a public. These processes are competition between media, depiction of identifiable scene and personae, revelation of conflicts and different outcomes, and the construction of a moral that audiences could adapt to their own lives. Chouliaraki's and Fine and White's models underline the transformation of external reality into media representation, which makes one or several features of an event more salient than the others. In disaster coverage, human suffering is presented more prominently than other aspects of the event (Moeller 1999; Houston et al. 2012). Human suffering depicted on media representation becomes an "ethical issue" (Boltanski 2004) because, at least in theory, there is an opportunity to alter the circumstances. Faced with a representation of suffering, an individual audience asks themselves what good decision they should make in response to it.

Needless to say, this formulation as an ethical issue is not the only possibility, nor is there only one formulation made by different groups, civil society organizations, local government agencies, and media. On an individual level, the problem can be formulated as a political issue by focusing on how the authorities take responsibilities and distribute resources. Or it can be formulated as an economic problem with focus on the business lost to the event. These said, studies show that media audiences feel that ethical problem formulation is the most authentic (Vrasti and Montsion 2014 via Pantti 2015) as it is "independent of formal institutions or the ideology of market forces" (Pantti 2015, p. 624).

On the organizational level, the problem is a little bit different since on this level audiences see themselves as collective actors in relation to other organizations and groups. The problem is no longer practical intervention but strategic effort to achieve their goal by working with, or competing with, other organizations (Saputro 2016). Jalin Merapi's effort was very illustrative of this issue. The key organizers of Jalin Merapi saw that the problem was disconnection between the needs of the local community and the potential helpers from the outside. Regarding this, mass media could not help them because they were interested in their own issues and they trusted the officials of government agencies more than the local people to get an overview of the circumstances. The contrast between how the people's media formulated the problem and that of the mainstream media can be seen in the early days of the eruption. Whereas national media were interested in the scale of explosion and whether a respected spiritual guardian of the volcano died in the explosion, Jalin Merapi brought attention to the specific needs of the people living on the slopes of Mount Merapi who were least exposed by the media. In so doing, Jalin

Merapi rode on the surging attention of media audience, which was brought about by converging attention of different media outlets, and at the same time attempted to direct their attention away from the issues raised by mass media. Further, when media established themselves as middle-person between audiences and the refugees through their fund raising and relief distribution efforts, Jalin Merapi took a different path. Jalin Merapi saw it as a problem of disconnection and localised information that required media to match specific, local needs of the people of Merapi with what audiences could provide and deliver.

Imagining the Others

Since media reach across different geography and socio-cultural groups, audiences can be living in a different world from the disaster-affected people, sometimes widely apart (von Engelhardt 2015). Although they might live close physically, the fact that the audiences do not have to evacuate their homes separates them from the precarious existence that the disaster-affected people live. In other words, the distance between media audience and the suffering others can assume three meanings: geographic, socio-cultural, and experiential. It was the audiences' separation from the refugees in experiential term that limits their capacity for imagination, as has been proposed by von Engelhardt (2015, p. 700). However, their inability to imagine the feeling of someone in that position, which limits their ability to empathize, is not relevant to their willingness to help (pp. 698–9) because willingness to help does not require ability to feel other's emotion.

Based on my study of the volunteers, donors, and those who complained, I would propose that although imagining the suffering other's emotional state is not necessary for audiences' willing to act, they still need to imagine them. Many times it was out of practical reason. For example, I observed a group of donors having a hard time deciding whether they should buy teabags or something else since they did not know what the refugees preferred, considering their taste and limited amenities. Similar imagination work had to be performed by a cook in the city who provided ready-made meals for the refugees in the rural area; she had to imagine their preferences every time she cooked. This kind of imagination work is different and goes beyond a unity of feeling, which means "how much of ourselves we see in a suffering other" (Cialdini et al. 1997 via von Engelhardt 2015, p. 699). Imagination here concerns how we put ourselves in a suffering other's position as a basis for making practical decisions. If they cannot overcome this barrier, their action would not be fully materialized or be as successful.

Another level of imagination was performed by volunteers of Jalin Merapi when they reframed victims as survivors (*penyintas*) to deliberately shift the discourse of passivity to active engagements and equality between external helpers and the suffering others. This is in direct opposition with mainstream media's discourse, which used the more common term "victims" (*korban*), which Jalin Merapi activists reserved only for lives lost. The word survivors was used to facilitate other audiences to imagine the suffering others as being equal and capable of coping with the impact. However, I should note here that the term has not been widely accepted and I have not assessed how the general audience received the reframe.

Building Resources

Resources in a disaster response situation are a real issue. In fact, since the daily routine of a region is disrupted, aid is needed from the outside. Money and in-kind donations need to be collected first, and then distributed later. My interviews show that most often audiences relied on their personal, work-related, and residential networks to solicit donations. Sometimes what they managed to collect was perceived to be too little that they had to redo the collection or hold a special event to raise funds. In many cases, an organization had a special reserve of funds (*dana sosial*) that they normally used when a member fell sick or passed away, yet its use cannot be immediate since the members needed to agree on it first. Further, decision about an organization's collective fund was usually made in a normal meeting, which can be held weekly, bi-weekly, or monthly, which delayed the delivery of the greatly needed funds.

The problem of resources even manifested in the less material type of collective action, namely making complaints to the Indonesia Broadcasting Commission through its online system. The data from the Indonesia Broadcasting Commission shows that the profile of those who complained (dominantly male) differed from the usual profile of gossip-show viewers (dominantly female). I would argue this means that the means to complaint was not evenly distributed among the viewers, with males having more resources than females. In this case, the means to complaint are access to the Internet and personal computer (because the online form was not designed for mobile devices). In other words, access to resources was a barrier to the audiences.

Organizations that organized volunteers also faced problem of resources. Combine Resource Institute, who managed the Jalin Merapi volunteers, could not provide the funds to prolong their efforts although some refugee centers still needed them. The volunteers themselves gradually left the project since most of them were students and their universities started to resume their activities. This shows how resources become barriers at the individual and collective levels.

Taking Action

Each form of collective action has a specific mechanism, which puts constraint on and can become barriers to audiences who want to join the action. The barriers are not specifically related to media but pertain to how a collective action is organized to achieve its goal. In my study, the principle mechanism of collective donation through media is *conversion of ownership*. The money to be donated had been available before, whether it belonged to an individual or an organization. What happened in the donation was a series of decisions to convert the ownership, from an individual to an organization, from an organization to the media, and from media to the people of Mount Merapi. It was a move from private to (semi)public ownership. This mechanism creates a barrier because media preferred money than in-kind donations that some groups felt small money was not worth donating. Volunteering revolved around *doing* relief-related activities on location. Those who individually helped gathering and publishing information on social media felt that what they did was a lesser form of volunteerism since it did not involve going to the refugee centers or disaster-affected areas. It is understandable that young people were dominant in the volunteering initiatives. Lastly, filing complaints was basically attempts to *transmit* grievances using official channels. Those who did not master the know-how of transmitting the message, or those who complained outside official channels, were not recognized and were not acted on by the authorities.

Necessary Sociotechnical Conditions for (Resilient) Audience-Driven Relief Response

Having learned how collective attention transforms into collective action, we are ready now to examine sociotechnical conditions that make this possible. The conditions are of hybrid nature since it was clear from the example that social organization alone could not prompt people to volunteer, or to donate, or to complain at that scale and velocity. Synchronic action at such scale and velocity was not possible to organize due to the geographically dispersed nature of media audiences and their fragmented attention to diverse media outlets and issues. Hence, information and communication technologies were indispensable in synchronizing their attention and action. Conversely, mere information and communication technologies could not select which disaster to attend to, or decide on the forms of action in response to a specific facet of a disaster event. It was the compound of technical and social arrangements that solve both synchronicity and intentional problems.

These two aspects, technical aspect of synchronicity and social aspect of intention, coincided to what Amir and Kant (2018) called informational relations aspect where "information is organized and managed to support continued operation of sociotechnical systems" (p. 13). The Mount Merapi eruption in 2010 radically changed the operation of the Jalin Merapi network, from local monitoring of the mountain for their own communities to connecting communities on different sides of Mount Merapi with each other and with outside communities. To do so, they recruited information volunteers (Saputro 2016) to report the needs of far larger communities and built an ad hoc media center (Mendonca et al. 2007). The cooperation between outside volunteers, outside donors, and local communities was made possible by organizational and communication networks that were built by the community radio stations and local NGOs in 2006. Another condition that made such cooperation possible was the rise of smart phones that made reporting, accessing, and distributing information possible for general media users. In 2010, approximately, thirty million people in the island of Java were internet users, whose majority accessed it through their mobile phones. Hence, the rapid adoption of social media by internet users in Java was a necessary condition for the exceptionally participatory disaster relief response.

From a limited collaboration between community radio stations, the Jalin Merapi Network was greatly expanded in 2010 to receive information from different sources (local community radio, hundreds of volunteers, and thousands of media users) and published them on various platforms (crater monitoring on livestream of handheld transceiver, list of needs on live online document (Google), continual stream of social media update (Twitter), and self-produced news articles) that all can be accessed on its website. Multiple media platforms and a variety of contents were a deliberate choice by Jalin Merapi to bring together different groups of people with different media usage. Merapi people commonly used handheld transceivers and short message services, whereas the urban young people used smartphones and other computational devices. Their command on various media platforms fell in line with their strategy to frame the problem of relief response as a disconnection between potential helpers and survivors. Furthermore, they organized information about survivors' need using the demand-driven approach, by publishing the need so that potential helpers can respond to it rather than publishing supply that might result in conflict over short supply, as they had observed in disaster response in 2006.

The choice of media platform is important in such critical times as certain media platforms enable some groups of people while constraining some others at the same time. For example, in the case of complaints to the Indonesian Broadcasting Commission about doomsday scaremongering on a national television channel, those who complained were largely congruent with internet users (male, young, educated) rather than television viewers because the complaints were accepted through online forms on its website. A sample of their complaints slightly showed their biased view toward the suffering others as being easily panicked based on their abstract knowledge of them.

When building resources for collective action, properly designed sociotechnical arrangements are crucial to achieving its goal since sociotechnical problems in disaster relief response, such as asynchronous input and delayed feedback, can create information redundancies that result in overrepresentation of some needs and underrepresentation of some other. It in turn leads to oversupply of certain items and undersupply of some others. Furthermore, sociotechnical arrangements in this particular case created a gatekeeper problem where gatekeepers (donation organizers and distributors) selected the kind of donation being accepted and distributed, which might be less congruent with the survivors' needs. Different from this, Jalin Merapi provided the right condition for participation in which the donors could directly connect with and assess the needs of the survivors.

The final sociotechnical conditions for participatory disaster relief response were two sides of the same thing, namely the publicness of the issue and action. It cannot be emphasized enough that in disaster individuals and groups from the surrounding area come to help (Drabek and McEntire 2003); yet the scale and efficacy of such actions are left unexamined. My case study revealed that potential helpers monitored the circumstances and their social circles when considering whether or not they would join a collective action. Consequently, it mattered to them whether an action was responded to by many people and that the problem itself affected many people. Continual report from mass media and people's media (Jalin Merapi), on the one hand, and continual monitoring by media users through their traditional media consumption devices and smart phones, on the other hand, assured that these two conditions were met.

It should be noted, however, that all of what is written here is based on a study of one particular disaster and we need to do more studies of positive cases of resilient communities to learn how different sociotechnical arrangements work for them. Yet, the goal is not to map an exhaustive model of resilient informational relations, but to distill several core models that work. A pragmatic approach is appropriate in this case since disaster is not an everyday occasion that is easy to come by and, more importantly, learning something that works can save lives.

Conclusion

Convergence of people to disaster-affected area should be integrated into the framework of resilience because it has been found in all disasters that it becomes part of the disaster response itself. A more severe form of disaster will usually invite even bigger convergence (Quarantelli 2003, 2008). The convergence can create a new problem, ranging from traffic congestion, delay of logistics, confusion of information, or becoming a safety hazard itself. It can also solve some of the problems during and right after the disaster event, such as evacuating, search and rescue, collecting and distributing aid, rapid damage assessments, and refugee centre operation, among others. In such context, understanding how audiences respond to mediated disaster events and how each process of their progression toward collective action is faced with barriers will allow us to identify these barriers and how we can reduce and overcome them.

Description of how convergence arises from media audience perspective also serves to put audiences' agenda into consideration, without reducing them into results of public relation stunts of the media and government (Carah and Louw 2012; Saputro 2016). By taking into account audiences' agenda, we can develop a participatory disaster response framework between government agencies and the public (IFRC 2011) and we also realize that media have roles other than as information producers, namely as solidarity makers and providers of emotional support (Hindman and Coyle 1999; Perez-Lugo 2004). Needless to say, media audiences' perspective will not and cannot serve to integrate other perspectives of disaster response since it only concerns the use and the users of media outlets. On the contrary, we cannot lose the opportunity to integrate media user's behaviours into our disaster response framework.

References

Adams, P. (2012). *Grouped: how small groups of friends are the key to influence on the social web.* New Rider.

Amir, S. & Kant, V. (2018). Sociotechnical Resilience: A Preliminary Concept. Risk Analysis, 38: 8–16. https://doi.org/10.1111/risa.12816.

Boltanski, L. (2004). *Distant suffering: morality, media, and politics.* Cambridge University Press.

boyd, d. (2011). Social network sites as networked publics: affordances, dynamics, and implications. In *Networked self: identity, community, and culture on social network sites*, edited by Z. Papacharissi, 39–58. Routledge.

Bruns, A. & Burgess, J. E. (2012). Local and global responses to disaster: #eqnz and the christchurch earthquake. In *Conference proceedings of disaster and emergency management conference*, edited by P. Sugg, 86–103. Brisbane, AST Management.

Bruns, A., Burgess, J., Crawford, K., & Shaw, F. (2012). *#qldfloods and @qpsmedia: crisis communication on Twitter in the 2011 south-east Queensland floods.* ARC Centre of Excellence for Creative Industries and Innovation.

Carah, N. & Louw, E. (2012). 'Inundatedy by the audience: journalism, audience participation, and the 2011 Brisbane flood'. *Media International Australia* 144(1), 137–145.

Cedillos, V. & Alexander N. (2010). 'Learning from earthquakes: Indonesia earthquake and tsunami'. *EERI Newsletter* 44(12), 4.

Chouliaraki, L. (2006). *The spectatorship of suffering.* Sage.

Cialdini, R. B., Brwon, S. L., Lewis, B. P., Luce, C., & Neuber, S. L. (1997). 'Reinterpreting the empathy-altruism relationship'. *Journal of Personality and Social Psychology* 73(3), 481–494.

Couldry, N., Livingstone, S., & Markham, T. (2007). *Media consumption and public engagement: beyond the presumption of attention.* Palgrave Macmillan.

Dayan, D. (2009). 'Sharing and showing: television as monstration'. *Annals of the American Academy of Political and Social Science* 625(1), 19–31.

Drabek, T. T. & McEntire, D. A. (2003). 'Emergent phenomena and the sociology of disaster: lessons, trends, and opportunities from the research literature.' *Disaster Prevention and Management* 12(2), 97–112.

Ewart, J. & Dekker, S. (2013). 'Radio, someone still loves you!: talkback radio and community emergence during disasters'. *Journal of Media and Cultural Studies* 27(3), 365–381.

Fine, G. A. & White R. D. (2002). 'Creating collective attention in the public domain: human interest narratives and the rescue of Floyd Collins'. *Social Forces* 81(1), 57–85.

Fritz, C. E. & Williams, H. B. (1957). 'The human being in disasters: a research perspective'. *Annals of the American Academy of Political and Social Science* 309(1), 42–51.

Hindman, D. B. & Coyle, K. (1999). 'Audience orientations to local radio coverage of a natural disaster'. *Journal of Radio and Audio Media* 6(1), 8–26.

Houston, J. B., Pfefferbaum, B., & Rosenholtz, C. E. (2012). 'Disaster news: framing and frame changing in coverage of major U.S. natural disasters, 2000–2010'. *Journalism & Mass Communication Quarterly* 89(4), 606–623.

Houston, J. B., Hawthorne, J., Perreault, M. F., Park, E. H., Hode, M. G., Halliwell, M. R., McGowen, S. E. T., Davis, R., Vaid, S., McElderry, J. A., & Griffith, S. A. (2014). 'Social media and disasters: a functional framework for social media use in disaster planning, response, and research'. *Disasters* 39(1), 1–22.

IFRC. (2011). *Protect. Promote. Recognize: volunteering in emergencies.* Report. Swedish Red Cross.

Knight, A. (2006). 'Covering disasters and the media mandate'. *Media Asia* 33(1–2), 47–71.

Kurnia, A. (2014). 'Surono berdiri di cincin api.' Available at: http://geomagz.geologi.esdm.go.id/surono-berdiri-di-cincin-api. Accessed at June 18, 2017.

Meier, P. & Munro, R. (2010). 'The unprecedented role of SMS in disaster response: learning from Haiti'. *SAIS Review* 30(2), 91–103.

Mendonca, D., Jefferson, T., & Harrald, J. (2007). 'Collaborative adhocracies and mix-and-match technologies in emergency management: using the emergent interoperability approach to address unanticipated contingencies during emergency response.' *Communication of the ACM* 50(3), 45–49.

Moeller, S. D. (1999). *Compassion fatigue: how the media sell disease, famine, war and death.* Routledge.

Nielsen (2010). Watching natural disaster on tv news. *Nielsen Newsletter* 11(Nov), 1. Jakarta, Nielsen.

Nugroho, Y., Amalia, D., Nugraha, L. K., Putri, D. A., Tanaya, J., & Laksmi, S. (2013). Creating content, shaping society: do Indonesian media uphold the principle of citizenship? *Engaging media, empowering society: Assessing media policy and governance in Indonesia through the lens of citizens' rights.* Jakarta, CIPG and HIVOS.

Pantti, M. (2015). 'Grassroots humanitarianism on Youtube: ordinary fundraisers, unlikely donors, and global solidarity'. *The International Communication Gazette* 77(7), 622–636.

Perez-lugo, M. (2004). 'Media uses in disaster situations: a new focus on the impact phase'. *Sociological Inquiry* 74(2), 210–225.

Quarantelli, E. L. (1994). Emergent behaviors and groups in the crisis time period of disasters. *Preliminary paper #206.* Newark, Disaster Research Center.

Quarantelli, E. L. (2003). A half century of social science disaster research: selected major findings and their applicability. *Preliminary paper #36.* [online]. Available at: http://dspace.udel.edu:8080/dspace/handle/19716/297. Accessed June 29, 2017.

Quarantelli, E. L. (2008). 'Conventional beliefs and counterintuitive realities'. *Social Research: An International Quarterly* 75(3), 973–904.

Saputro, K. A. (2014). Engaged audiences in the mediated disaster of Mount Merapi in 2010. Thesis. Sheffield Hallam University.

Saputro, K. A. (2016). 'Information volunteers' strategies in crisis communication: the case of Mt. Merapi eruption in Indonesia 2010'. *International Journal of Disaster Resilience in the Built Environment* 7(1), 63–72.

Starbird, K. & Stamberger, J. (2010). Tweak the tweet: leveraging microblogging proliferation with a prescriptive syntax to support citizen reporting. In: *Proceedings of the 7th International ISCRAM Conference, May 2–5.* Seattle, ISCRAM.

Starbird, K. (2011). Digital volunteerism during disaster: crowdsourcing information processing. In: *CHI 2011, May 7–12.* Vancouver, ACM.

Surono, Jousset P., Pallister, J., Boichu, M., Buongiorno, M. F., Budisantoso, A., Costa, F., Andreastuti, S., Prata, F., Schneider, D., Clarisse, L., Humaida, H., Sumarti, S., Bignami, C., Griswold, J., Carn, S., Oppenheimer, C., & Lavigne, F. (2012). 'The 2010 explosive eruption of Java's Merapi volcano: A '100-year' event.' *Journal of Volcanonology and Geothermal Research* 241–242(1), 121–135.

Tanesia, A. & Habibi, Z. (2011). 'JALIN Merapi community information system in response to Mount Merapi's 2010 eruption. n.p. Accessed at http://www.amarc.org/documents/Caribbean_Conference/CR_ResponseJALIN_MerapiEruption_EN.pdf

von Engelhardt, J. (2015). 'Studying western audiences vis-a-vis mediated distant suffering: a call to venture beyond media studies'. *The International Communication Gazette* 77(7), 695–707.

Vrasti, W. & Montsion, J. M. (2014). 'No good deed goes unrewarded'. *Global Society* 28(3), 336–355.

Webster, J. G. (1986). 'Audience behavior in the new media environment'. *Journal of Communication* 36(3), 77–91.
Webster, J. G. (2002). 'The internet audience: web use as mass behavior'. *Journal of Broadcasting & Electronic Media* 46(1), 1–12.
Webster, J. G. (2005). 'Beneath the veneer of fragmentation: television audience polarization in a multichannel world'. *Journal of Communication* 55(2), 366–382.
Webster, J. G., Phalen, P. F., & Lichty L. W. (2006). *Ratings analysis: the theory and practice of audience research.* Lawrence Erlbaum Associates.

PART IV

Engineered Systems

CHAPTER 9

Post-Fukushima Controversy on SPEEDI System: Contested Imaginary of Real-Time Simulation Technology for Emergency Radiation Protection

Shin-etsu Sugawara and Kohta Juraku

Introduction

Once the striking news on the crisis at TEPCO's Fukushima Dai-ichi Nuclear Power Station (hereafter Fukushima Dai-ichi NPS) arrived, public attention was focused on the magnitude, trend, and consequence of off-site radioactivity release. Everyone in the area surrounding the reactors wanted to have concrete information to protect them from its possible harm. In this context, a simulation system attracted very strong public attention during the emergency period of the accident. It was called SPEEDI (System for Prediction of Environmental Emergency Dose

S. Sugawara (✉)
Central Research Institute of Electric Power Industry, Tokyo, Japan
e-mail: sugawara@criepi.denken.or.jp

K. Juraku
Tokyo Denki University, Tokyo, Japan

S. Amir (ed.), *The Sociotechnical Constitution of Resilience*,
https://doi.org/10.1007/978-981-10-8509-3_9

Information). SPEEDI is a Japanese original real-time simulation technology for emergency radiation protection developed and implemented since the 1980s. It was expected to produce both graphical and quantitative simulation results, which would enable the relevant actors to take appropriate actions to protect the people from undesirable radiation exposures.

However, as this chapter will illustrate in the following sections, it was not the case that SPEEDI achieved results in line with such expectations. The insufficiency of its performance has been repeatedly criticized, especially concerning the public disclosure of its results, and its handling during the acute emergency period of the disaster. It became one of the controversial points of off-site nuclear emergency management. Evaluations on this topic were divided even among major accident reports. For example, the national governmental accident report argued that more active use of SPEEDI could help better decisions, while the National Diet's report counter-argued that it could not due to many technical limitations.

Many academic papers, including some by Science and Technology Studies (STS) scholars, supported the argument of the governmental commission's accident report: it could be used to help lower exposure to radiation for the evacuees, but intentionally and/or structurally, SPEEDI's results were concealed by the relevant governmental organizations for some reasons, such as to save face, bureaucrats' collective fear against "panic" (so-called elite panic), and so on (for example, see Matsumoto 2013a; Shimazono 2016; Tsuneishi 2012). They criticized the secrecy of the government based on the assumption that SPEEDI did provide useful scientific predictions and could have helped the actual radiation protection measures taken during the acute phase of the nuclear disaster. We believe, however, this problem should be deeper than an issue of openness vis-à-vis secrecy. It's not a problem of information disclosure in crisis, but rather, a problem relevant to the nature of computer simulation technology and its public imaginary, and their complicated and tight interaction at the interface of science, technology, and society. We need to problematize the assumption–that SPEEDI itself is useful, the problem is bureaucracy–in a critical manner, relativizing the popular story centering on the well-known anti-disaster simulation system.

To explore our research concerns, we conducted qualitative research including document surveys and 13 semi-structured interviews for 24 informants including the original developers of SPEEDI, national/local governmental officials, domestic/international experts of nuclear safety and emergency management, and so on. Description and analysis in this chapter is based on those qualitative research data.

This chapter is organized as follows. First, we will give an overview of SPEEDI and its role and function in the nuclear emergency management framework in Japan as determined before the Fukushima nuclear accident, including the institutional structure, the basic specifications of SPEEDI, and the history of institutionalization of this system. The following section will describe the "failure" of SPEEDI in the first response to the Fukushima nuclear accident, as well as how this was evaluated and problematized by the relevant investigation reports. In the third section, we will deeply analyze the reasons of this discrepancy by illustrating the contested views towards SPEEDI among relevant actors. Finally, we will discuss the effect of the public imaginary on these kinds of real-time simulation technologies, which are believed to make society more "disaster-proof" through "scientific prediction." The similarities and gaps between these beliefs and the experts' notion on certainty/uncertainty of SPEEDI's output and its possible usage will be touched on. Interactions and consequences of these notions centering on "resilience" will be examined critically.

SPEEDI and Framework of Nuclear Emergency Response in Japan Before 2011

SPEEDI (System for Prediction of Environmental Emergency Dose Information) is a system designed to simulate and predict the dispersion of radionuclides and its radiological consequences in nuclear emergencies. This simulation technology, in combination with topographical data, can visually show the geographical distribution of released radioactive materials, and the extent of possible resulting land contamination or human radiation exposure. In addition, this system had a function of connecting the organizations relevant to nuclear emergency management, including national and local agencies, through leased lines. With these technical features, SPEEDI was expected to support emergency response in nuclear accidents.

Before depicting the roles of SPEEDI, this section will explain the framework of nuclear emergency response in Japan before the Fukushima accident. Figure 9.1 illustrates the outline of the institutional framework for nuclear emergency management.

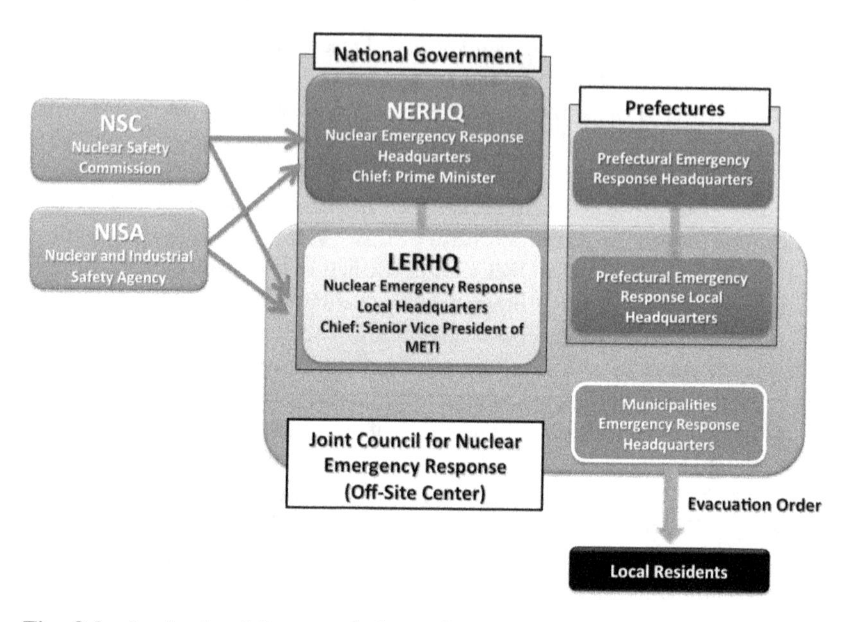

Fig. 9.1 Institutional framework for nuclear emergency management in Japan. (Cited and modified from Nuclear Emergency Response Headquarters Government of Japan 2011)

In this framework, whenever a precursor of a severe accident was observed, the operators of commercial nuclear reactors were supposed to report it immediately to the METI (Ministry of Economy, Trade and Industry) and relevant local governments. If deemed necessary, the prime minister would declare a "Nuclear Emergency Situation," and then establish the National Emergency Response Head Quarters (NERHQ) and the Local Emergency Response Head Quarters (LERHQ). The NERHQ, who would be headed by the prime minister, consisted of the regulatory body (NISA: Nuclear and Industry Safety Agency[1]), the advisory body for nuclear safety (NSC: Nuclear Safety Commission), the technical support organization (JNES: Japan Nuclear Energy Safety Organization), and other relevant agencies such as the National Police Agency and the Fire and Disaster Management Agency.[2] The LERHQ, whose chief is the senior vice minister of the METI, would be set up at the Off-Site Center, a command center at the disaster area, located several kilometers away from nuclear power plants in most sites. The prefectures and municipalities would also establish their emergency response headquarters at the Off-Site Center. Those national and local headquarters, connected with each other via a teleconference system, would form the Joint Council for Nuclear Emergency Response, enabling relevant organs to share information immediately and collaborate closely.

The Head Quarters would consider necessary actions and measures in light of site information and expert advice. As deemed necessary, the head of the NERHQ would decide to take protective actions such as stay-in-house instructions and evacuation orders, which would be delivered and implemented to the subject area via the municipal mayors.[3] It was when the Head Quarters judged the necessity and scope of protective actions that SPEEDI had been expected to play a central role (Fig. 9.2).

Before 2011, two computer simulations had been expected to play a central role in emergency decision-making: ERSS and SPEEDI. ERSS, standing for the Emergency Response Support System, is a simulation system to predict accident progress by analyzing reactor information transmitted from the operators. This system, originally developed by the NUPEC (Nuclear Power Engineering Corporation[4]) after the Chernobyl accident, was designed to calculate and predict abnormal conditions of the

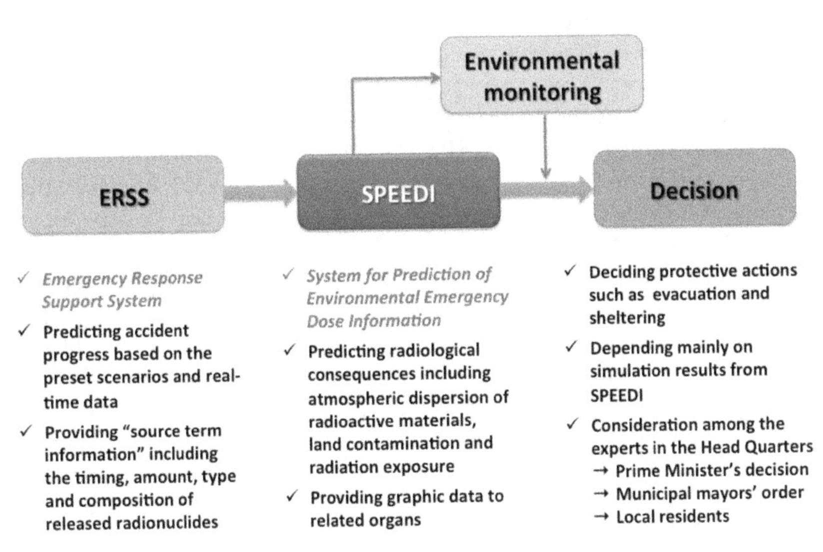

Fig. 9.2 Decision-making flow in case of nuclear emergency. (Cited and modified from Sugawara 2015)

reactor based on the pre-set scenarios and real-time data transmission from nuclear reactors, for example, when the temperature of nuclear fuel reaches 1200 degrees Celsius, or when the pressure in the primary containment vessel is up to three times higher than designated. In the emergency response framework, ERSS was expected to provide the so-called source term information, which includes the timing, amount, type and composition of radionuclides that would be released from the damaged nuclear reactors.

Having received real-time weather data from the AMeDAS (Automated Meteorological Data Acquisition System),[5] SPEEDI would calculate near-future climate conditions such as wind direction, wind power, and rainfall, taking into consideration specific topographical data around the accident site. Combining these weather projections with the source term information from ERSS, SPEEDI would simulate the atmospheric dispersion of radioactive materials and the possible radiation dose in the surrounding area. The calculation time was claimed to be approximately 15 minutes (when predicting for six hours) between receiving orders from the relevant organs to transmitting the calculation results—it could actually be called a "real-time" simulation.

Then, SPEEDI would provide the graphical data representing the prediction of possible accident consequences to the relevant organizations including the Head Quarters, regulatory agency, related ministries, and local governments. The expert team of the Head Quarters would then deliberate the plans for protective areas, taking into consideration many factors such as the output from SPEEDI, environmental monitoring data, meteorological information, and so on. On a practical level, however, the decisions on evacuation and stay-in-house orders were heavily dependent on SPEEDI's prediction.

Figure 9.3 is a typical example of a geographical chart provided by SPEEDI and the protective action areas that were set in a past nuclear emergency drill. The two overlapping ovals stretching from the northwest to the southeast delineate SPEEDI's calculated values of possible radiation dose. The evacuation and stay-in-house zones are shown as a concentric area of the nuclear power station and as an additional area downwind in three directions, which are clearly based on SPEEDI's simulation. In this case, the decision of evacuation did not receive influence from actual measured values from environmental monitoring–this was quite natural

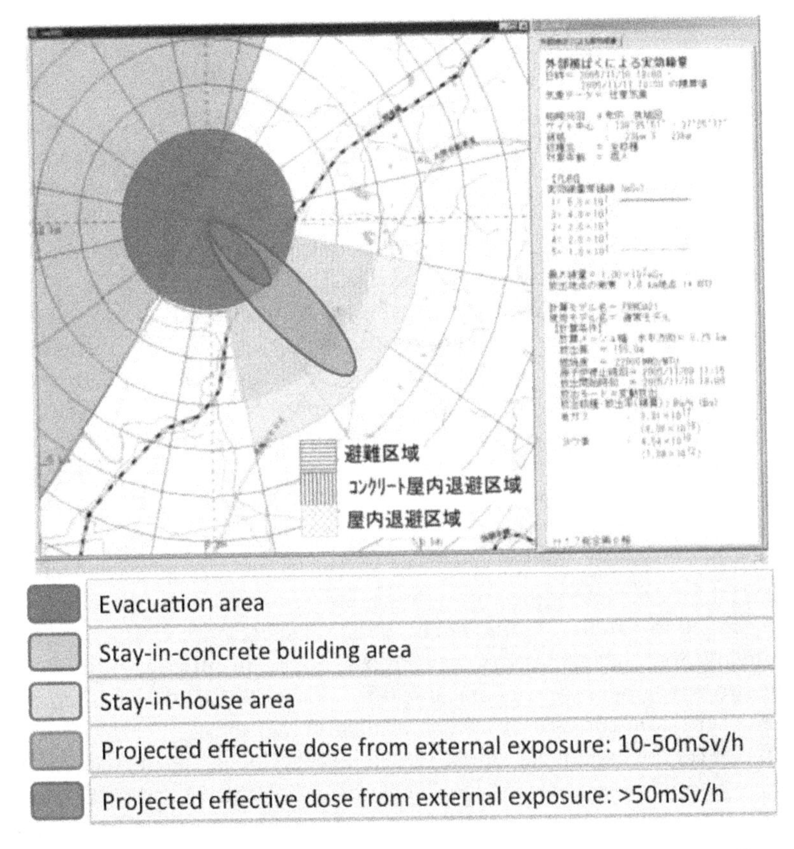

Fig. 9.3 An example of SPEEDI results and evacuation areas in the past disaster drills [in Niigata prefecture for TEPCO's Kashiwazaki-Kariwa Nuclear Power Station]. (Cited from Nuclear Safety Commission of Japan 2006)

because in this exercise, the accident was assumed to have been mitigated successfully before any release of radioactive materials, an assumption common to many accident scenarios in pre-Fukushima disaster drills.

As depicted, SPEEDI outputs were regarded as "scientific evidence" and were directly referred to in the decisions on protective actions, which could be extremely vital for protecting the lives of local people.

History of SPEEDI Research, Development, and Institutional Implementation

The origin of SPEEDI can be traced back to the early 1980s. In reaction to the Three Mile Island accident, the Central Disaster Prevention Council and the NSC recognized the necessity of developing simulation technology to support nuclear emergency management.

From 1980 to 1985, the Japan Atomic Energy Research Institute (JAERI)[6] developed the first version of SPEEDI by looking at the ARAC (Atmospheric Release Advisory Capability) project in the United States. Originally, SPEEDI did not have the capability of predicting near-term climate conditions but only the function of analyzing present weather status (Imai et al. 1985). In 1986, the STA (Science & Technology Agency[7]) started to entrust the Nuclear Safety Technology Center (NUSTEC), a public interest corporation under the Ministry of Education, Science and Culture (MEXT),[8] with the pilot use of SPEEDI and the development of networks among the relevant ministries and local governments. After the Chernobyl accident, which occurred in the same year, JAERI utilized SPEEDI for calculating the atmospheric dispersion of radionuclide plumes and verified its validity (Chino et al. 1986). In 1992, the NSC revised the Guideline for Nuclear Emergency Preparedness, an internal rule of the NSC, to clarify necessary technical features where SPEEDI had taken the place of the conventional diffusion calculation methods.

Based on the experience of the JCO critical accident, Japan enacted the Act on Special Measures Concerning Nuclear Emergency Preparedness in 1999. The utilization of SPEEDI was officially institutionalized in the Nuclear Emergency Response Manual that concretely stipulates the roles of the relevant governmental ministries and agencies for realizing the legislative purposes of the Act. In this manual, the outputs from SPEEDI simulation was formally mentioned as one of the information for deciding evacuation plans together with emergency monitoring data. As the years went by, prediction calculation of this system acquired a more prominent position as "basic information for selecting monitoring points and for deciding areas subject to evacuation orders etc." (NSCJ 2008). Since then, the framework for utilizing SPEEDI has been established and strengthened; in normal situations, the NUSTEC executes regular calculations in a predetermined manner, while in emergency situations, it operates the system in accordance with the orders from the —— and related ministries. NUSTEC and its subcontractor regarded SPEEDI as "an important system pertaining to the fundamentals of nuclear emergency preparedness in Japan" (Misawa and Nagamori 2008), and focused on developing and maintaining their calculation capabilities in case an accident should happen.

SPEEDI has been well appreciated, not only by administrative practitioners, but also from an academic standpoint. In 2008, this system was awarded the "Atomic Energy Historic Award" by the Atomic Energy Society of Japan. Upon receiving this Award, one of the developers of SPEEDI made remarks as follows: "SPEEDI is adopted by the government as the central system for national nuclear disaster preparedness so it has contributed to the social acceptance of nuclear technology..." (Atomic Energy Society of Japan 2009). This was the best time for SPEEDI.

Realities of Nuclear Emergency: Case of Fukushima Nuclear Accident

When the Fukushima accident happened, did SPEEDI play its expected role as explained above? Unfortunately, the answer is NO. A series of evacuation and stay-in-house orders from the national government, issued six times from March 12 to March 15, 2011, were decided without support from SPEEDI. The situation at that time was as follows (Fig. 9.4).

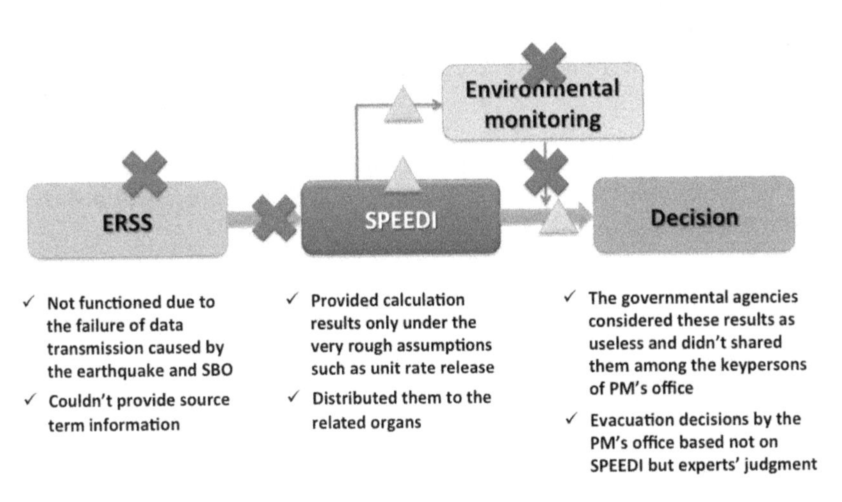

Fig. 9.4 Decision-making flow in the case of the Fukushima accident. (Cited and modified from Sugawara 2015)

In the first place, ERSS could not provide source term information due to the failure of data transmission caused by the earthquake and the SBO (Station Black Out) triggered by the tsunami, which made it impossible to get precise plant data. The lack of source term information was a big obstacle for SPEEDI in implementing predictive calculations as it was expected to. The NUSTEC, the operator of SPEEDI, tried to calculate with the conditions that the MEXT and relevant agencies ordered, for example, "unit rate release," an assumption of releasing 1Becquerel per hour (Bq/h) for both noble gas and iodine, and the predetermined data of a "hypothetical accident," defined as "an accident thought to never occur from an engineering perspective" by the Regulatory Guide for Reviewing Reactor Site Evaluation.

The following figures are some typical examples of the results of SPEEDI calculation in the initial stage of the Fukushima accident.

The contour lines in the chart represent possible dispersion trajectories in the case where 1Bq of radioactive iodine would be released every hour from Unit 1 of the Fukushima-Daiichi NPS on March 12. It is instantly noticeable that the diffusion patterns are highly dependent on the timing of a major release: the direction of dispersing radionuclides would shift seaward around noon, westward and then northwestward in the evening, in response to the predicted wind direction and power.[9]

These calculation results were supposed to be transmitted immediately to the Head Quarters and shared among the relevant organs. Hundreds of figures, indeed, were distributed to the MEXT, METI, NSC, NISA, and so on. Nevertheless, the key personnel in the prime minister's office who had a practical role in deciding evacuation orders had not been informed about these simulation outputs. Investigation Committee on the Accident at the Fukushima Nuclear Power Stations (ICANPS) report described this issue as follows:

> Cabinet Secretariat staff, who received the predictions from NISA staff before dawn on the morning of March 12, treated them as reference information and did not report to Prime Minister Naoto Kan (hereinafter referred to as "Prime Minister Kan"). Also NISA itself did not report the predictions to Prime Minister Kan either. (ICANPS 2011: 298)[10]

According to a testimony by Tetsuro Fukuyama, the Deputy Chief Cabinet Secretary at that time, the politicians in the PM's office were not aware of the existence of SPEEDI until March 16 or 17 (ICANPS 2012:

65). This resulted in the decision-making regarding evacuation and sheltering in the initial response being completely ignorant of the calculation results of SPEEDI, contrary to the original design concept of the emergency management framework.

Issue of Secrecy/Openness

As described, SPEEDI kept providing simulation results when the accident occurred–although its calculations could not reflect the actual details of accident progress. The governmental agencies that received those results, such as the MEXT, NSC, and the NISA emergency response center, thought that "those simulation outputs without input of source term information were totally useless." That is why they didn't proactively share the results with the executives in the PM's office or disclose them to the public.[11]

SPEEDI suddenly started to gain public attention on March 15. Dr. Ryugo Hayano, professor at the Department of Physics in the University of Tokyo, an individual drawing much attention during the accident while nuclear engineering professors were losing their public trust, tweeted about SPEEDI around 14:20 on this day. Subsequently, a journalist demanded the disclosure of the outputs from SPEEDI and WSPEEDI, a worldwide version of SPEEDI, at a MEXT press conference on the same day. The official at MEXT responded by stating that it would be "under deliberation."

On the March 16, during a meeting at the PM's office, roles on the operation of emergency monitoring were divided into two: to implement environmental monitoring and disclose its results would be the role of MEXT, and to assess monitoring data would be that of NSC. In the afternoon, the SPEEDI operators at NUSTEC who were stationed at MEXT moved to the NSC secretariat. With this move, the jurisdiction of SPEEDI was also substantially transferred to the NSC.[12]

Under enormous social pressure for the disclosure of SPEEDI calculation results, the first to be publicized was the NSC's reverse estimate calculation result on March 23.

The figure on the left represents the aerial environmental monitoring data provided by US-DOE (Department of Energy) and MEXT disclosed on May 6, and the figure on the right is the reverse calculation results by SPEEDI. At first glance, the contour lines extending northwestward illustrated in both figures bear a strong resemblance. It may seem that SPEEDI can predict radiological consequences with great accuracy. In reality, however, the right-hand figure is not a prediction made in the midst of the accident, but a back-cast simulation (so-called reverse calculation) based on environmental monitoring data, calculated one week after the occurrence of the accident. Nonetheless, it was reported as if it were a prediction, and led to public "misunderstandings" pointed out in the National Diet of Japan Fukushima Nuclear Accident Independent Investigation Commission (NAIIC) report, that is, the view that the government intentionally concealed valuable simulation results despite SPEEDI providing them in a timely manner. Following the instruction on April 25 from Yukio Edano, Chief Cabinet Secretary at that time, the NSC, MEXT, and NISA disclosed their SPEEDI data, respectively. Although almost all the results were publicly open by the end of April, Goshi Hosono, then Special Adviser to the Prime Minister, announced in a press conference on May 2 that there were still some results that remained unreleased.

The whole story of disclosure deepened public suspicions and reinforced the frame of understanding that problematizes SPEEDI as an issue of secrecy/openness rather than that of the capacity and limitations of simulation technology.

Continuing Controversy

Today, the controversy over SPEEDI continues. The two most official accident investigation reports on the Fukushima accident, ICANPS and NAIIC, provide an almost opposite evaluation, as shown in Table 9.1. ICANPS argued that it could be used for further optimizing evacuation measures even if it could provide only hypothetical calculation results using unit rate release assumption. It criticizes the lack of flexible and adoptive handling of SPEEDI by relevant actors. On the contrary, NAIIC concluded that it could not be used effectively due to an intrinsic limitation of SPEEDI's function resulting from the lack of source term information, which will be explained later in this chapter. It clearly rejects the opinion that SPEEDI could have been used for improving the evacuation procedure even if unit rate release assumptions were made and criticizes this idea as a total "misunderstanding." These contrasting views regarding SPEEDI are all the more notable given that NAIIC retains a more critical attitude through most of the report rather than ICANPS.

Dichotomous disputes between these two major positions have continued since then. The Nuclear Regulatory Authority (NRA), the regulatory body newly established after the Fukushima accident, tried to settle this

Table 9.1 Comparison of ICANPS and NAIIC reports concerning the evaluation of SPEEDI

ICANPS report[a]	*NAIIC report*[b]
... And yet, even when source term information cannot be obtained, SPEEDI can obtain the results of predictions for the unit amount of discharge (1Bq/h), and actually did so. This means that if unit quantity emission predictions had been provided, the various municipal bodies and residents would probably have been able to decide which evacuation routes and directions were most suitable for them, taking advantage of their firsthand knowledge of local road conditions	... ERSS and SPEEDI are systems to forecast future events based on a certain calculation model. In particular, if release source information cannot be retrieved from ERSS, SPEEDI data alone lacks the accuracy to serve as a basis for establishing evacuation zones. In this accident, events unfolded very rapidly and the results of the projection could not be utilized for the initial evacuation orders
...The fact that SPEEDI was not used effectively indicates that it did not occur to any of the competent authorities that the system could serve a role in evacuation, and that no clear-cut decision had been made in advance regarding which other organization should take over the role of the local NERHQ at the Off-site Center when it lost its functionality to provide information to the public	...After the accident, release source information could not be retrieved from ERSS for many hours. Related organizations, including NISA and MEXT, concluded that SPEEDI's calculated results could not be utilized, and so the system's results did not contribute to the initial evacuation orders. The results of the calculations from reverse estimate calculations that were disclosed by NSC at a later date were misunderstood, and believed to have been projections from the time the accident occurred. This gave rise to further misunderstanding and the belief that the government could have prevented residents' exposure to radiation had the results been disclosed promptly and SPEEDI been effectively utilized in making decisions about the initial evacuation orders

[a]ICANPS (2011), Chapter VII

[b]NAIIC (2012), Chapter IV

controversy recently after a long internal deliberation process on the nuclear disaster management planning. They issued a statement on "Principles on Protective Actions in a Nuclear Disaster Event" (title tentatively translated by the authors) in March 16, 2016 and clearly declared that "the calculation results of diffusion simulation has no reliability" in the event of a nuclear emergency. The statement denies the advantages of using SPEEDI for emergency response and strongly criticizes the possible adverse effects of prediction-based evacuation optimization because "it is impossible to change the destination or route of evacuation once the evacuation has been put to action" (NRA 2016). This portrayal was a total denial of the use of SPEEDI in the eyes of public opinion.

However, just five days before the issuance of NRA's statement, the Ministerial Council for Nuclear Power Utilization decided on a document that includes pro-SPEEDI anti-nuclear disaster recommendations, titled "Strategy for Improvement of Anti-Nuclear Disaster Measures" (title tentatively translated by the authors). It says that the national government shall never prevent the prefectural/municipal governments from referring to SPEEDI calculation results at their own responsibility, both for their nuclear emergency decision-making and emergency drills (Cabinet Secretariat of Japan 2016). This document is sub-titled "Learning the lessons of Fukushima and responding to the recommendations by the National Governors' Association" and suggests that this is a reactive endorsement by the national government responding to the bottom-up request to appropriate the advantages of SPEEDI in an emergency. The principle suggested by the Ministerial Council's report was also confirmed in 2017 by another report by the Inter-Ministerial Liaison Group for Counter Measures to Nuclear Disasters. It is a clear signal that the government still recognizes an "at your own risk" usage of SPEEDI by local governments, knowing that NRA's judgment is against this. Although this dispute is not yet settled, SPEEDI is no longer embedded in the official procedures for disaster management.

Perception Gap in the Uncertainties of SPEEDI

Where are such dichotomous views coming from? To address this question, the authors did bibliography surveys and interviewed a series of relevant actors, including the developer of SPEEDI, national and local governmental officials, the experts in the fields of nuclear safety and emergency management, and so on. Analyzing our investigation carefully, the authors collated the views on the usability and the roles of SPEEDI held by each actor as well as the past regulatory guidelines and disaster drills (Table 9.2).

It can be clearly seen that the recognitions of SPEEDI are vastly different, ranging from "useful without hesitation" to "useless at all." The

Table 9.2 Perception gaps regarding SPEEDI

Actors	*View towards SPEEDI*	*Conclusion*
Developer of SPEEDI	One of the reference materials for emergency management experts. Not expected to be disclosed directly to the public.	*USEFUL* with some conditions
Officials of nuclear hosting local gov'ts.	Important basic information for decision-making to protect their people. Use the outputs for considering emergency monitoring and/or evacuation orders along with other information.	*USEFUL* with some conditions
Ex-guideline for nuclear disaster response	One of the sources of information for making evacuation decision with emergency monitoring results. But, not so clearly defined.	*USEFUL* without careful thoughts
Residents of nuclear siting areas	Expect the output to be critical information to avoid any additional radiation exposure. The problem in the Fukushima case was secrecy (not technical limitations).	*USEFUL* from an innocent sense of expectation
Customary practice in the past disaster drills	Substantial information for decision-making as 'scientific evidence'. Sharing SPEEDI outputs among relevant organs, local gov., etc.	*USEFUL* without hesitation
Some experts in nuclear safety and emergency management	No one can predict accurately when and how a nuclear reactor will lose its confinement function. There is need for departure from prediction-oriented decision-making style.	*USELESS* at all

authors, based on a qualitative interview analysis, found that this recognition gap is derived from the different views on the uncertainties of SPEEDI prediction.

There are three main uncertainties accompanying the calculation process of SPEEDI: (1) the uncertainty of source term information, (2) that of the weather forecast, and (3) the uncertainty that comes from modeling phenomena such as atmospheric dispersion, soil deposition, and radiation exposure of the released radioactive materials. The most conflicting point between pro-SPEEDI experts and anti-SPEEDI experts lies in how to address the uncertainty regarding source term information. As described earlier, the emergency response framework before the Fukushima accident expected to use the advantages of SPEEDI for deciding evacuation orders

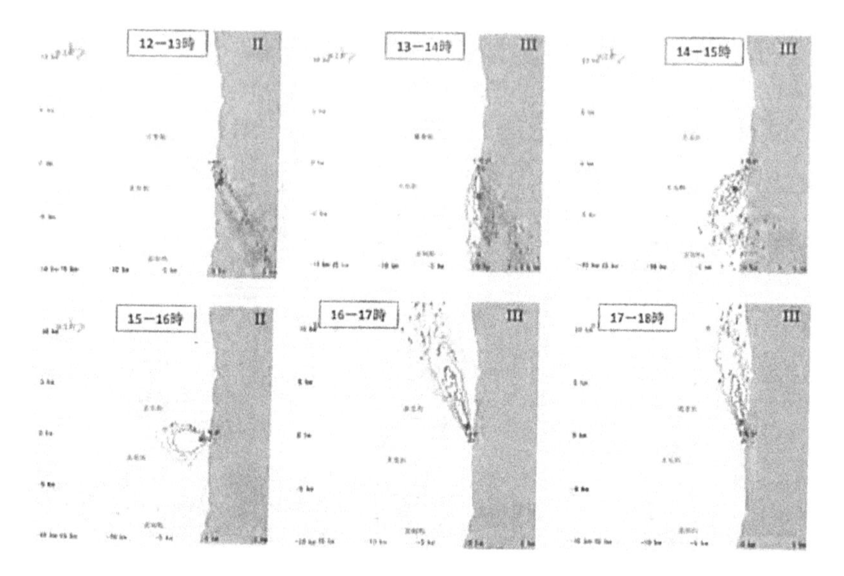

Fig. 9.5 Examples of the SPEEDI calculation results in the initial response of the Fukushima accident. (Cited from a MEXT's report (MEXT 2012). Calculation conditions are as follows: March 12, 2011, prediction of equivalent dose, Iodine, 1 year child, Unit 1, Fukushima-Daiichi NPS)

on the presumption that ERSS could provide source term information even in emergencies. Yet, in the acute phase of this accident, ERSS did not work well, and SPEEDI could only make predictive calculations under very rough assumptions such as unit rate release conditions, as depicted in Fig. 9.5. The controversy among the experts centers on whether such rough simulation results calculated without source term information would contribute to making better decisions or not.

The pro-SPEEDI experts such as atmospheric diffusion model experts think that even if we cannot obtain source term information, we can operate SPEEDI and make calculations by entering some value into the system. Regardless of the emission volume, mode of emission, chemical property of released radionuclide, and so on, the illustrated dispersion pattern at a specific point in time would be basically the same, because the predicted weather status and geographical conditions are the same (Sato 2013). The validity of prediction models for simulating local weather and

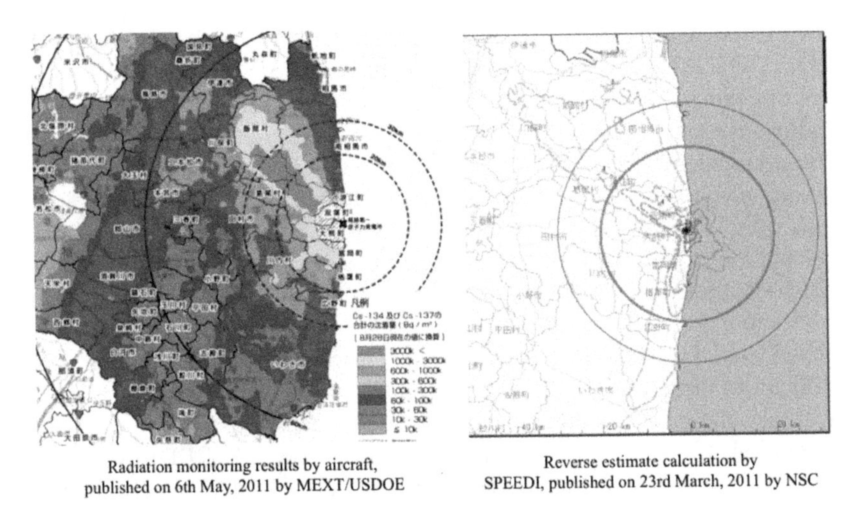

Fig. 9.6 Comparison with the aerial environmental monitoring data and the reverse calculation results by SPEEDI. (Cited from MEXT 2011a, b)

diffusion conditions is quite high, which has been demonstrated by the consistency between the environmental monitoring results and the reverse calculation estimates as shown in Fig. 9.6. Hence, the simulation results of SPEEDI, even without source term information, could help decision-makers consider the direction and timing of evacuation.[13]

Contrary to this, anti-SPEEDI experts such as a part of nuclear safety and emergency management experts counter those arguments as follows: never let it be said that the simulation results without source term information can be useful for deciding protective actions. With calculation under very rough assumptions, judging the necessity of sheltering and evacuation by comparing the absolute value of the predicted radiation dose to the predetermined dose criteria[14] at any point is invalid, despite this being customary practice in past disaster response exercises. In the first place, there is no way we can accurately predict the release of radioactive materials from damaged nuclear reactors and obtain source term information for SPEEDI under emergency situations. When used as planned, ERSS would tell us when the temperature and pressure of the primary containment vessel would be two or three times higher than the designed ones. It is impossible, however, to estimate when the vessel will break and lose its containment function in reality. Also, it is truly difficult to forecast which part of the vessel would give way and how it would break, which heavily influences the volume and composition of released radionuclides. Moreover, as we experienced in Fukushima, the release of radioactive materials could be reoccurring and intermittent depending on the accident progression. Despite the fact that the dynamic change of emission volume is also vital for estimating human radiation exposure, it is also quite difficult to predict the former with accuracy. Therefore, the calculation results without source term information never make a substantial contribution to decide evacuation orders.[15]

Some experts of nuclear emergency management have more negative views: even if we can get source term information in a timely manner, it is difficult to make use of SPEEDI for optimizing protective measures when you take the time and effort needed for real evacuation actions into consideration. If people began to evacuate in response to SPEEDI output at a certain point, wind direction would change easily in the midst of evacuation. What's more, simultaneous evacuation of thousands of people in their own cars may result in heavy traffic congestion as the evacuees in Fukushima actually experienced. It is not realistic to change the direction of evacuation in the middle of such social turmoil.[16]

When we look closely, however, the actors who foster high expectations for SPEEDI, such as the developers of this technology, are positive about using SPEEDI only under certain conditions. For example, one of the SPEEDI developers stressed that its simulation results should not be "primary information" as the governmental manuals regarded but simply "one of the reference information sources." The developers didn't think that SPEEDI could illustrate accurate scientific predictions and semi-automatically determine appropriate political decisions, even before the Fukushima accident.[17] Such an understanding of the developers of this system, who are supposed to be well versed in the capability, limitations, and uncertainties of this technology because they themselves produced it, had not been shared widely in the process of institutionalizing nuclear emergency management, which also applies to the post-accidental dichotomous controversy over SPEEDI.

On the other hand, there is still high expectation for SPEEDI held by some local governments and people living in nuclear hosting areas, far away from the strongholds of technology production activities.[18] Such anticipation seems to be based not only on optimistic views of the uncertainties of prediction but also upon an innocent and earnest hope that SPEEDI will protect them from unwanted radiation risks through its scientific capability. We postulate that this kind of public imaginary on simulation technologies has been formed by past public relations and the name of this simulation system itself: the name "SPEEDI" could easily encourage the public to expect that the government and experts would handle everything well and quickly by using this prominent technology. Also, more fundamentally, it is likely that such public attitudes have a commonality with risk-avoiding behaviors in everyday life based on "scientific prediction," such as weather forecasts, which could be one of the sociotechnical constitutions deeply embedded in our contemporary risk society.

Contested Imaginary of Real-Time Simulation Technology for Emergency Radiation Protection

The authors found a sign of effect of the public imaginary on these kinds of real-time simulation technologies, which are believed to make society more "disaster-proof" through "scientific prediction." Voices come from outside the community of core experts on nuclear safety engineering and disaster mitigation for complex systems, which call for the increased use of SPEEDI during emergencies: the criticism regarding the calculation result disclosure in the acute phase of the Fukushima accident, the ICANPS arguments in their report, and the prefectural governors' request. The public and non-core-professionals set their expectations high on the benefit of so-called real-time simulation technology for emergency response. On the other hand, core professionals have always been wary of their technology being used as a "convenient" tool for emergency decision-making and public protection. They were not involved in the policy process to promote the application of SPEEDI for "real-time-optimization of counter measures" in such a naive manner. This gap in the perception of the utility of SPEEDI can be interpreted as a variant of the "certainty trough" effect articulated by MacKenzie (1990). We can call it a "certainty zigzag" after the shape of the curve as shown in Fig. 9.7.

The vertical axis indicates the degree of perceived uncertainty of SPEEDI, while the relevant stakeholders are plotted on the horizontal axis. Actors on the right-hand side of the chart support the pro-SPEEDI arguments that call for further utilization of it in the acute phase of a nuclear disaster. Institutional designers of nuclear disaster preparedness, disaster drill practitioners, and local government officials, all of them have strong interests to be in favor of that theory, because it would give them

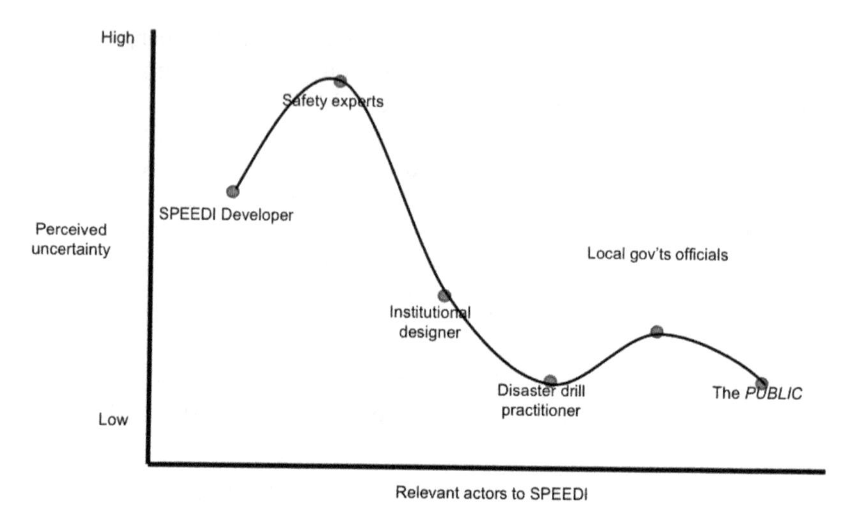

Fig. 9.7 Certainty zigzag

strong and politically "neutral," easily justifiable guidance for their decisions on nuclear disaster management. Of course, these people are influential in the policy process of nuclear disaster management in general.

On the other hand, nuclear safety engineering experts, who have even stronger influence in the newly established regulatory body, the NRA, estimate the uncertainty of SPEEDI calculation the highest. They are skeptical against the entire concept of real-time simulation-informed emergency decision-making because of their own insights on the reliability of source-term calculation based on real-time reactor behavior simulation like the ERSS. They are very cautious towards the development of a popular story that invites a too high expectation on the predictability of its outputs. Although some of these experts had tried before the Fukushima accident to change this kind of prediction-oriented emergency response scheme by making recommendations through political channels and academic societies such as the Science Council of Japan (SCJ 2005), their arguments did not have enough political mobilization to change the basic concepts. After the post-3/11 regulatory reform, their presence in the regulatory body has become larger, in comparison to how they had been marginalized before the Fukushima accident. The ban of the use of SPEEDI for nuclear emergency decision-making is one of the symbolic pivotal points of the new regulatory policy, which could work as a leverage to demonstrate this vivid change. Actually, they have criticized the SPEEDI R&D project for a long time as a not promising and too costly one although they belonged to the same institution with the SPEEDI team, JAEA (Japan Atomic Energy Agency, formerly JAERI: Japan Atomic Energy Research Institute).[19]

The SPEEDI developers, who have backgrounds in the scientific field rather than engineering, have been excluded from the policy process of nuclear disaster preparedness consistently. Their mission in mind is solely "scientific R&D" of quantitative computer modeling, not practical design and operation of SPEEDI as an administrative system. In fact, operation and implementation of SPEEDI has been managed by clearly different institutions, with governmental divisions handling the policy process, while the NUSTEC, a satellite organization of MEXT (formerly STA), took charge of operation. They of course have strong interests to support pro-SPEEDI administrative planning to legitimate the funding investment for their research, but it means they have not had a clear incentive to let the SPEEDI supporter stakeholders know about the limitations and reliability of their simulation technology, which could undermine beliefs shared among the actors. Such a social and political environment structurally neglected the most reliable and detailed experts on SPEEDI from thinking about the appropriate way to use the advanced simulation technology. The structural exclusion of the most relevant expertise is very similar to the process illustrated by McKenzie.

Epilogue: Contested Imaginary and Sociotechnical Resilience

If we would discuss the background of this phenomena further, scientific optimism and progressivism in general may be a relevant factor, or it might be something to do with a relatively higher consciousness of improvements in disaster prevention measures shared by Japanese people who live in a rather natural-disaster-prone geography. It could be a result of Japanese people's self-portrayal as a "technologically advanced" nation. It could just be a simple expression of a traditional enthusiasm for prophecy, too. In any case, this kind of deep-rooted public imaginary for "anti-disaster capacity improvement through the advancement of simulation technology" has affected the trajectory of public controversy and has had a considerable effect on social decision-making regarding disaster prevention and resilience of society (and relevant organizations) under emergency situations. However, it should be noted that such a simple and well-meant will shared by citizens could lead to ineffective, infeasible anti-disaster plans as this chapter has shown. In this light, the contested imaginary of real-time simulation technology should be a focus when we discuss the sociotechnical constitution of resilience. Both the public imaginary that advanced techno-science would bring a better society, and the scientists' belief that they should always strive to improve technologies are not necessarily harmful in themselves. Nevertheless, once those two innocent prepositions are entangled with each other and come to function as a hybrid "sociomaterial structure," which Amir and Kant (2018) defined and highlighted as one of the main features of sociotechnical resilience. Over-dependence on simulation technologies becomes institutionalized in the emergency management framework, with a great deal of financial and human resources devoted to its development and maintenance. Thus, the public imaginary would serve as the entity that underpins political power in this structure, maintaining and strengthening its power uncritically along the way.

While "anticipatory practices," another theoretical component of sociotechnical resilience as formulated by Amir and Kant in the same paper, is often depicted as highly desirable for minimizing disaster risks, it should be noted that an anticipatory culture shared among relevant actors does not always guarantee an effective disaster response as the case of SPEEDI shows. Anti-disaster culture and related practices formed among the public and a confined group of experts and practitioners, which are undefined but therefore solid, may have a weak impetus for reflecting on their inherent problems in light of criticism from experts outside their imagined community. Such a lack of mechanism for rigorous identification and readjustment of shortcomings in the prevailing system, a typical feature of "structural disaster" as elaborated by Matsumoto (2002=2012, 2012, 2013a, b), could be recognized as a negative aspect of resilient sociotechnical system. This reflexive insight may open chances for STS and disaster studies to elaborate and cultivate the concept of resilience, which tend to be conveniently and arbitrarily referred to in a non-critical manner.

It is indeed inevitable that advanced simulation technology brings positive imaginaries and high social expectancy, which could promote even more development of science and technology. We believe, however, that it should be the role of sociotechnical risk scholars to identify and address inherent problems accompanied by well-painted dreams of sociotechnical resilience before something terrible happens again.

Acknowledgment Part of this chapter was supported by the "Social Scientific Research Support Program" of Tokai Village, Ibaraki Prefecture, Japan. Also, it was partly supported by JSPS Kakenhi 17K18139 "Critical Analysis of Dysfunction of Real-time Disaster Damage Simulation System: "SPEEDI" and Other Cases." The authors would like to extend their sincere gratitude for these public funding support for academic research. The authors greatly appreciate the help of Rin Watanabe, a graduate student at the Department of Nuclear Engineering and Management, the University of Tokyo, as a research assistant for the JSPS Kakenhi project.

Notes

1. NISA was to be a special organ under the METI.
2. NISA, NSC, and JNES; all of these organizations no longer exist after the post-Fukushima accident nuclear governance reform.
3. In Japan, the primary responsibility and authority to decide evacuation orders based on the Basic Act on Disaster Control Measures belongs to municipalities. However, lessons learned after the JCO criticality accident in 1999 (nuclear disaster response requires advanced expertise, affected areas often cross the borders of municipalities and call for integrated and coordinated actions, and so on) prompted the subsequently established Act on Special Measures Concerning Nuclear Emergency Preparedness to provide a legal framework for the national government to take part more actively in nuclear disaster response.
4. Predecessor of JNES.
5. AMeDAS is the national network of weather observation that was developed, established, and operated by the Japan Meteorological Agency (JMA). It has about 1300 monitoring points and provides real-time data of precipitation, air temperature, duration of sunshine, wind direction and speed, and snow depth in snowfall areas. Each monitoring post sends these data to the AMeDAS center at the headquarters of JMA every ten minutes. The data is used as the basic data for JMA's weather forecast and other activities. It started its operation in November 1974. The acronym AMeDAS stands for a Japanese sentence 「雨出す」 (Ame-Dasu: Providing Rain (data)).
6. Principal national research institute for scientific and engineering research in the nuclear field in Japan. Predecessor of JAEA.
7. STA used to be the "control tower" of nuclear research and development in the national government at that time. It was merged with the Ministry of Education, Science and Culture by the 2001 Central Government Reform.

8. As mentioned in the previous endnote, it was reformed into the Ministry of Education, Culture, Sports, Science and Technology (MEXT) in 2001.
9. Technically speaking, not only the timing of release, but also the "mode" is critically important for calculating the diffusion of radioactive particles and radiation exposure dose rates. For example, the distribution of contamination could be very different in the following two cases: (a) very quick, instant big release of radioactive particles caused by some explosive event at reactor, and (b) prolonged, trickling release due to some leakage from the gaps and joints of reactor vessels. If influential weather conditions (such as wind direction and speed and/or precipitation) change in the middle of the simulated time period of (b), the distribution of the contamination could be very different from (a).
10. In the case of the Fukushima nuclear accident, the PM's office made directly important decisions on taking protective actions with support from the experts of NSC and some temporary political appointees, which was not in accordance with the prearranged plan described by the guidelines and manuals. This was because the disaster was a combination of a huge earthquake and tsunami, and the Off-Site Center didn't function well because of the heavy damage at relevant local governments due to these natural disasters.
11. A similar thing happened in Fukushima. While the Fukushima prefectural government had received SPEEDI calculation results since the evening of March 11, the secretariat of its emergency response headquarters did not share them among the prefectural office, but erased them because the person in charge at that time thought that "utilizing SPEEDI results for deciding protective actions was not the roles of the prefecture" (Fukushima Prefectural Emergency Response Headquarters Secretariat 2012).
12. The details and backgrounds of this re-distributing of the roles of environmental monitoring and SPEEDI are depicted in the accident investigation report (ICANPS 2011, Chapter V). The dominant view is that MEXT had forced the responsibility of SPEEDI to the NSC. This has incited severe critiques such as "this case was the most hideous of Kasumigaseki's (bureaucratic) battle for shifting blame onto each other" (Funabashi 2012: 389).
13. Interviews with an aerial dispersion simulation expert in Japan (national research institute), December 27, 2016, two aerial dispersion simulation experts in Japan (national research institute), November 17, 2016, and three aerial dispersion simulation experts in Japan (private research institute), February 8, 2017.
14. Evacuation is stipulated when the external exposure effective dose is above 50mSv, and sheltering is stipulated at 10–50mSv.

15. Interview with a nuclear safety and probabilistic risk assessment engineering expert in Japan (former member of national research institute), September 30, 2016.
16. Interview with a nuclear safety and disaster prevention engineering expert in Japan (national research institute), October 30, 2016.
17. Interview with an aerial dispersion simulation expert in Japan (national research institute), December 27, 2016.
18. Interviews with a local municipal government officer in charge of disaster preparedness in Japan, November 16, 2016, former chairperson of a local council for nuclear safety improvement and transparency in nuclear siting community in Japan, January 30, 2017, five local prefectural government officers in charge of nuclear disaster preparedness in Japan, January 30, 2017, and, three local prefectural government officers in charge of technical implementation of nuclear disaster preparedness and radiation monitoring in Japan, March 21, 2017.
19. Abe (2014) and Nuclear Systems Association (2014) are the examples of such criticism made by nuclear safety engineering experts in Japan.

References

Abe, K. (2014), 「原子力安全と規制(2)福島第一事故の反省と原子力安全基盤の再構築」 (Nuclear Safety and Regulation (2) Reflection of Fukushima Dai-ichi Accident and Reconstruction of Nuclear Safety), Proceedings of the 2nd Summer Seminar of Nuclear Safety Subcommittee, the Atomic Energy Society of Japan. (in Japanese)

Amir, S. and V. Kant (2018), Sociotechnical Resilience: A Preliminary Concept. Risk Analysis, 38: 8–16. https://doi.org/10.1111/risa.12816.

Atomic Energy Society of Japan (2009), 『第1回(平成20年度)原子力歴史構築賞』(Atomic Energy Historic Award 2008), Atomic Energy Society of Japan. (in Japanese)

Cabinet Secretariat of Japan (2016), 「原子力関係閣僚会議決定 原子力災害対策充実に向けた考え方」 (“Decision of the Ministerial Council for Nuclear Power Utilization: Strategy for Improvement of Anti-Nuclear Disaster Measures”), March 11, 2016. (in Japanese)

Chino, M., H. Ishikawa, H. Yamazawa and S. Moriuchi (1986), “Application of the SPEEDI System to the Chernobyl Reactor Accident,” JAERI-M 86-142, JAERI M Report, Japan Atomic Energy Research Institute (October 1986).

Fukushima Prefectural Emergency Response Headquarters Secretariat (2012), 「福島第一原子力発電所事故発生当初の電子メールによるSPEEDI 試算結果の取扱い状況の確認結果」 (“Results of the Follow-up Survey on the Situation of SPEEDI Calculation Result Handling at the Early Stage of the Fukushima Dai-ichi Nuclear Accident”). (in Japanese)

Funabashi, Y. (2012), 『カウントダウン・メルトダウン(下)』 (*Countdown Meltdown*, Vol. 2), Bungei Syunju. (in Japanese)

Imai, K., M. Chino, H. Ishikawa, M. Kai, K. Asai, T. Homma, A. Hidaka, Y. Nakamura, T. Iijima and S. Moriuchi (1985), "SPEEDI: A Computer Code System for the Real-time Prediction of Radiation Dose to the Public due to an Accidental Release," JAERI 1297, JAERI Report, Japan Atomic Energy Research Institute (October 1985). Investigation Committee on the Accident at the Fukushima Nuclear Power Stations (ICANPS) (2011), "Interim Report" December 26, 2011.

Investigation Committee on the Accident at the Fukushima Nuclear Power Stations (ICANPS) (2012), 「聴取結果書」(福山哲郎民主党参議院議員、内閣官房副長官(当時)、事故対応全般について)2012年2月25日 ("Hearing Record" (to Mr. Tetsuro Fukuyama, a member of the House of Councilors of the Democratic Party of Japan, former Deputy Chief Cabinet Secretary, on the emergency responses to the Fukushima nuclear accident in general), February 25, 2012.

Mackenzie, D. (1990), *Inventing Accuracy: A Historical Sociology of Nuclear Missile Guidance*, Cambridge MA: MIT Press.

Matsumoto, M. (2002=2012), 『知の失敗と社会:科学技術はなぜ社会にとって問題か』 (The Structural Failure of the Science-Technology-Society Interface). Tokyo: Iwanami Publishing Co. (in Japanese)

Matsumoto, M. (2012), 『構造災:科学技術社会に潜む危機』(The Structural Disaster: Crisis Hidden in Techno-Scientific Society). Tokyo: Iwanami Publishing Co. (in Japanese)

Matsumoto, M. (2013a), 「構造災と責任帰属:制度化された不作為と事務局問題」 ("'Structural Disasters' Requiring Due Social Responsibility: Institutionalized Forbearance and Organizational Stakes"), *Journal of Environmental Sociology*, Vol. 19, pp. 20–44, Japan Association for Environmental Sociology, 2013. (in Japanese)

Matsumoto, M. (2013b), "'Structural Disaster' Long Before Fukushima: A Hidden Accident," *Development and Society*, Vol. 42(2), pp. 165–190.

Ministry of Education, Culture, Sports, Science and Technology (MEXT) (2011a), 「緊急時迅速放射能影響予測ネットワークシステム(SPEEDI)等による計算結果(平成23年3月)」 ("Calculation Results of System for Prediction of Environmental Emergency Dose Information (SPEEDI) (March 2011)"), Released in April 2011 (currently the data is succeeded and disclosed by the Nuclear Regulatory Authority, Japan). (in Japanese)

Ministry of Education, Culture, Sports, Science and Technology (MEXT) (2011b), 「文部科学省及び米国エネルギー省航空機による航空機モニタリングの測定結果について」 ("Airborne monitoring results by airplanes of MEXT and US Department of Energy") (currently the data is succeeded and disclosed by the Nuclear Regulatory Authority, Japan). (in Japanese)

Ministry of Education, Culture, Sports, Science and Technology (MEXT) (2012), 「東日本大震災からの復旧・復興に関する文部科学省の取組についての検証結果のまとめ(第二次報告書) 参考資料 SPEEDIの計算結果の活用・公表について」 ("Summary of the Result of Review on Measures for Recovery and Rebuild from the Great East Japan Earthquake Taken by MEXT (2nd Report) Appendix 'On the Use and Disclosure of SPEEDI Calculation Result'")

Misawa, M. and F. Nagamori (2008), 「緊急時迅速放射能影響予測(SPEEDI)ネットワークシステム」 ("System for prediction of environmental emergency dose information network system"), Fujitsu, 59(5), pp. 482–489. Fujitsu Ltd. (in Japanese)

Nuclear Emergency Response Headquarters Government of Japan (2011), Report of Japanese Government to the IAEA Ministerial Conference on Nuclear Safety: The Accident at TEPCO's Fukushima Nuclear Power Stations, June 2011.

Nuclear Regulatory Authority (2016), 「原子力災害発生時の防護措置の考え方」("Principles on Protective Actions in a Nuclear Disaster Event"), March 16, 2016. (in Japanese)

Nuclear Safety Commission of Japan (NSCJ) (2006), 「防災指針検討 WG 資料(平成18年8月2日)」 (Reference document for the working group on the Guideline for Nuclear Emergency Preparedness, August 2, 2006). (in Japanese)

Nuclear Safety Commission of Japan (NSCJ) (2008), 「環境放射線モニタリング指針」 (Guideline for Environmental Radiation Monitoring).

Nuclear Systems Association (2014), 『福島第一原子力発電所事故と原子力のリスク』 (Fukushima Dai-ichi Nuclear Accident and Risk of Nuclear Power), *NSA Commentary Series*, No. 21, Japan Atomic Industrial Forum. (in Japanese)

Sato, S. (2013), 『放射能拡散予測システムSPEEDI:なぜ活かされなかったか』 (*Prediction System for Radionuclides Diffusion SPEEDI: Why It Was Not Utilized*), Toyo Shoten. (in Japanese)

Science Council of Japan (2005), "Promotion of Research Activities on Prevention against and Recovery from Environmental Contamination by Radioactive Materials," Science Council of Japan. (in Japanese)

Shimazono, S. (2016), 「放射線健康影響をめぐる科学の信頼喪失——福島原発の初期被曝線量推計を中心に」 (Loss of trust for science on health effects of radiation exposure: Case of estimation of initial exposure dose of Fukushima nuclear accident), S. Shimazono, H. Goto, A. Sugita and T. Kobayashi (eds.) 『科学不信の現代を問う:福島原発災害後の科学と社会』 (*Analysis of Contemporary Society on Distrust for Science: Science and Society after the Fukushima Nuclear Disaster*), Godo Shuppan, Japan, 2016.

Sugawara, S. (2015), "Decoupling the cognitive link between information usefulness and disclosure: A case study on SPEEDI," presented at Society for Social Studies of Science 2015 Annual Meeting, Denver, Colorado, November 13, 2015.

The National Diet of Japan Fukushima Nuclear Accident Independent Investigation Commission (NAIIC) (2012), "The Official Report of Fukushima Nuclear Accident Independent Investigation Commission" July 5, 2012.

Tsuneishi, K. (2012), 「STSセンター:SPEEDIと地震予知」, *International Business and Management Forum*, Institute of International Business and Management, Kanagawa University, Vol. 23, pp. 141–148, 2012. (in Japanese)

CHAPTER 10

Saving Onagawa: Sociotechnical Resilience in the 3/11 Disaster

Makoto Takahashi and Masaharu Kitamura

Introduction

After the 2011 Great East Japan earthquake, the accident at the Fukushima Daiichi nuclear power plant, owned and operated by Tokyo Electric Power Company (TEPCO) , caused huge and tragic impacts on the people living not only in Fukushima but also on those living in nearby prefectures due to the uncontrolled release of radioactive materials. Published accident reports mainly focus on finding individual causes, responsible persons, and root causes. There is little emphasis on decisions and actions that reduced the severity of the tsunami impact on nuclear power plants (NPP) in the area. Some of the decisions made by the plant personnel on the site that contributed to mitigating the damages caused by the accident at Fukushima Daiichi have previously been analyzed (Takahashi and Kitamura 2013; Yoshizawa et al. 2015; 2016; Kitamura 2014, 2016). However, further investigations are necessary to learn the lessons to further improve safety

M. Takahashi (✉)
Tohoku University, Sendai, Japan
e-mail: makoto.takahashi@qse.tohoku.ac.jp

M. Kitamura
Research Institute for Technology Management Strategy (TeMS), Sendai, Japan

S. Amir (ed.), *The Sociotechnical Constitution of Resilience*,
https://doi.org/10.1007/978-981-10-8509-3_10

of nuclear power plants. In this chapter, we focus our attention on Onagawa NPP, which is owned and operated by TEPCO. Onagawa is the nearest nuclear power plant to the earthquake epicenter off the northeast coastline of Japan. It consists of three reactors, all of which were successfully shutdown and reached a stable condition while facing difficult situations similar to the Fukushima Daiichi nuclear power station. In this chapter, we analyze the case of Onagawa NPP based on the concept of resilience engineering combined with the concept of sociotechnical resilience. The basic concept in resilience engineering dictates that the mechanisms underlying failure and success are the same, and more attention is given to why things go well. However, in a more common view, it is easy to point out flaws in actions taken after the accident. In this chapter, we argue that positive contributions to mitigate accidents should also be emphasized to prevent recurrence. We believe that the Onagawa NPP was able to avoid nuclear disaster like the Fukushima Daiichi NPP because of the specific actions and situations; some of them were attributed to the inherent resilient behavior and some others were just by luck. In order to obtain lessons to prevent further disaster to happen, it is necessary to distinguish the inherent resilience behaviors from the serendipitous factors. This chapter highlights the contributing factors to reducing the disaster impact on Onagawa NPP. Furthermore, we separate those considered as the inherent features of organizational resilience from those by luck. The account of the critical situations at Onagawa during the 3/11[1] disaster in this chapter draws on published reports supported by interviews with the personnel in charge of safety emergency at Onagawa.

Resilience of Sociotechnical Systems

A nuclear power plant is in essence a sociotechnical system. It is a crucial infrastructure whose operation relies on the seamless integration between social and technical components. The interaction between these components determines the performance of the nuclear power plant as well as its ability to cope with a crisis. In this light, we would like to address resilience of nuclear power plants using two particular frameworks below in which technical and organizational aspects are regarded equally important. The first framework we refer to is "resilience engineering" developed by safety engineers, most notably Erik Hollnagel. Hollnagel et al. (2006, 2011) offer resilience engineering as a concept to enhance safety of sociotechnical systems where human beings are seen to play extremely important roles. It is worth noting that the term "engineering" in this concept does not necessarily means a narrow focus on technical features, as the concept is broadly defined to include organizational factors that contribute to system safety. The definition of safety in resilience engineering is the "ability to succeed under varying conditions." This is in contrast to a traditional view of "freedom from unacceptable risk" (ISO/IEC 2014). What distinguishes Hollnagel's resilience engineering from other safety science approaches is its strong emphasis on factors that go right than things that go wrong. This means, the framework stresses understanding the normal functioning of sociotechnical systems because it is through normal conditions we can learn how a sociotechnical system develops ability to withstand shock and crisis.

To further understand how resilience engineering is relevant to examining the success story of Onagawa, it is instructive to explain its core idea. As Hollnagel (2014) has elaborated, there are four underlying principles in recognizing resilient sociotechnical systems. First, systems and environments are always changing. In real-world situations, systems comprise people and organizations that change day to day following the nature of modern society. Second, most of the time important decisions are made based on imperfect information, which affect the efficacy of these decisions. This is a corollary of the previous principle. As changes within the boundaries of normal operations are not easily identified in advance, it is practically impossible to make ideal decisions. Thus, decisions are made based on limited amounts of information. Third, as systems are required to be beneficial and purse efficacy, it can result in "a drift to failure" unless adequate and special attention is given to maintain safety. In this issue, researchers of safety in other domains, such as medicine and aviation, recognize that a "drift to failure" is difficult to avoid unless special attention is given to maintain and improve safety. One remarkable example is the Fukushima Daiichi NPP in which the state of the plant had "drifted" toward the safety boundary and personnel in the plant were unaware of the current "distance" from this boundary. One possible reason is the biased attention given to address the problem of earthquakes and their countermeasures. TEPCO had previously experienced large earthquakes that hit another nuclear power station at Kashiwazaki-Kariwa in 2007. Since then, TEPCO had invested much time and effort to restart the plants, giving more attention to countermeasures against earthquakes until the Great East Japan earthquake (Hatamura 2012). Although this cognitive bias is a characteristic of human beings, it causes increasing difficulty in recognizing the "distance to failure."

Finally, while safety is of utmost importance, it is not the main purpose of a functioning system. This may sound different from the often-repeated stance in the industrial safety community stating, namely, "safety is always top priority in any situation." This seems to go against safety culture, where priority is given to the continuation of operations rather than to safety. However, as maintaining safety is a pre-condition of the continuation of systems operations, this view is consistent with placing priority on safety. It also means that safety is maintained with the aim of continuous system operations. Combined with the third principle, necessary measures to improve safety should not be avoided even if it results in the loss of short-term benefits and efficacy. In resilience engineering, this type of decision is known as "sacrificing judgment" (Hollnagel et al. 2006).

The core idea of resilience engineering lies in what Hollnagel (2014) terms as Safety-II. Resilience engineering is thus about how to engineer the Safety-II concept in order to enhance safety in the practical working environments. To understand the concept of Safety-II, to know the essence of resilience engineering, it is necessary to contrast it with Safety-I, which is defined as the way to enhance the level of safety by a *find and fix* strategy. Safety-I and Safety-II are not necessarily conflicting one another, but rather are complementary. Safety-II becomes effective based only on the condition that effort to enhance safety directed by Safety-I has already been performed. During the 1960s through 1970s, the concept of Safety-I had been effective enough when the level of complexity and interaction among systems was relatively low. It has become potentially ineffective for the sociotechnical systems where the level of interaction became higher and the coupling of systems became quite tight. One reason why Safety-I may not work well for advanced sociotechnical systems is that it is difficult and almost impossible to assume the strict cause-consequent relationships in the analysis of failure. It is assumed in Safety-I that all adverse outcome had identifiable causes that can be identified by following the cause-consequence relationship—the underlying principle of Root Cause Analysis (RCA). Although this may hold for the conventional mechanical entity, however complex and complicated it is, it may not hold for the sociotechnical systems, in which the mechanism of failure has emergent properties. This is the essential point to understand the Safety-II concept.

Another important concept in Safety-II is that things go right only when the appropriate adjustments have been performed by humans. Safety-I naturally assumes that things go well just by following the predefined procedures and if something goes wrong it is because there is something wrong. In other words, unless anything goes wrong, things will go well. In contrast to Safety-I, it is assumed in Safety-II that the success is achieved only when the careful adjustments of people are performed properly, although adjustments are always approximate and sometimes cause failures. Safety-II provides a new viewpoint to consider safety, but it doesn't provide operationally how to enhance the level of safety in the practical situations. The resilience engineering aims to provide the practical methodology to engineer the Safety-II concept to the practical sociotechnical systems, which is characterized by the four main potentials (Hollnagel 2017) to realize resilience as follows. First, it has potential to respond, which is indicated by knowing what to do and being able to respond to regular and irregular changes. Second, it has potential to monitor, which embodies in the knowledge of what to look for or being able to monitor those that affect or could affect the organizational performance, either positively or negatively, in the near term. Third, it has potential to learn, which means knowing what has happened and being able to learn from experiences, in particular to learn the right lessons from the right experiences. Finally, it has potential to anticipate uncertain developments further into the future.

In addition to resilience engineering, we apply the concept of sociotechnical resilience proposed by Sulfikar Amir and Vivek Kant (2018) to explain the sociotechnical features of the Onagawa NPP that contributed to saving the plant from a catastrophic ending. According to Amir and Kant, sociotechnical resilience is characterized by three main constituents, namely informational relations, sociomaterial structures, and anticipatory practices. They point out the importance of the conceptual theorization of what constitutes resilience for sociotechnical systems and also emphasize transformability as the main indicator of resilience. Informational relations refer to how information is organized and managed to support continued operation of sociotechnical systems. Sociomaterial structure refers to the mutual entanglement of human organization and material structures in sociotechnical systems. Anticipatory practice refers to the construction of regular activities aimed at anticipating possibilities of what would occur in the future.

Figure 10.1 illustrates the basic idea of resilience engineering based on the Safety-II concept. The particular point in resilience engineering is that it is an approach to consider seamless adjustment in the period of A and B in Fig. 10.1 assuming that variability of everyday performance and corresponding adjustments exist even during normal operation. Although the sociotechnical resilience view also takes anticipatory practice into account, it seems to have emphasis on the preparedness during regular activities to expect the possible event in the future. On the contrary, resilience engineering deals with the situation prior to, during, or following changes and disturbances, in which the four potentials play important roles to sustain required operations under both expected and unexpected conditions.

In this chapter, both the concepts of resilience engineering and sociotechnical resilience are employed as the basis of analysis to highlight the key features that allowed Onagawa to survive the unexpected events during the 3/11 disaster.

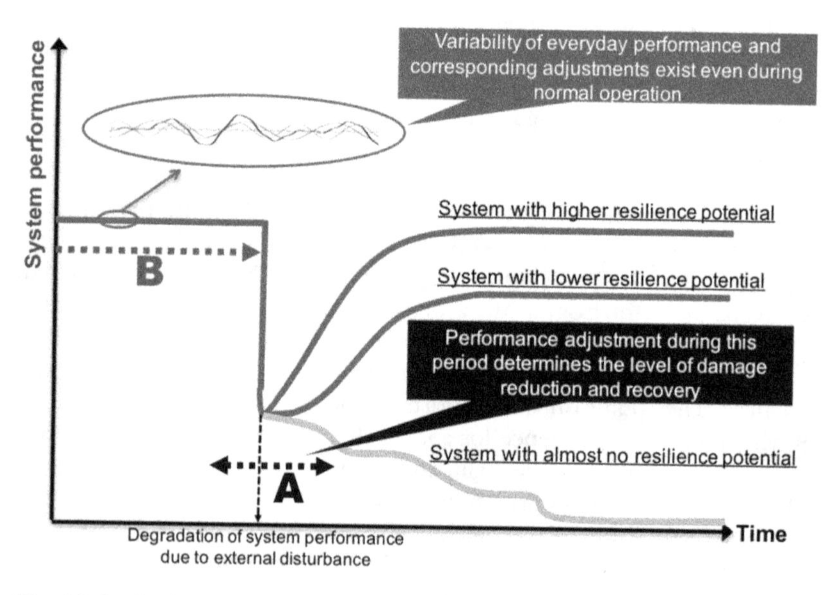

Fig. 10.1 Basic idea of resilience engineering based on Safety-II concept

Onagawa and Tsunami Risk

Onagawa nuclear power station is one of the two NPPs operated by Tohoku Electric Power Company.[2] It is located along the coastline of the Pacific Ocean, 70 km to the northeast of the city of Sendai as shown in the Fig. 10.2. It has three units and they are all boiling water reactors (BWR). The detail of each unit is shown in Table 10.1.

Fig. 10.2 Geographical location of Onagawa NPP

Table 10.1 Reactors of Onagawa NPP

Unit no.	*Capacity (MWe)*	*Commercial operation*	*Main constructor*
One	524	Jun-84	Toshiba
Two	825	Jul-95	Toshiba
Three	825	Jan-02	Toshiba/Hitachi

Historical account of this nuclear facility should be mentioned here in relation to the energy situation in Japan. In the 1970s, the Japanese economy grew rapidly thanks to the accelerated industrial policy. As Japan began exporting more industrial products overseas as the means for boosting high economic growth, the domestic demands for energy soared incredibly. During this time, the Japanese government saw nuclear energy as a viable source of energy. Thus, a number of nuclear power plants were scheduled to be built. Oil crises in 1973 and 1979 were another strong motivation to push the introduction of nuclear power forward (Okamoto 2017). Onagawa NPP was one of the first plants to be built. The construction of Onagawa started in 1979 and it was the first NPP operated by Tohoku Electric Power Company. By then, all of the reactor units at Fukushima Daiichi NPP had already become operational in 1979 and domestic companies such as Toshiba and Hitachi had developed capacity to build nuclear reactors based on their own technological resources. Although the Tohoku area didn't have a large industrial complex that required large amounts of electricity, Onagawa NPP was expected to play a role as a backup electrical energy source for the Tokyo metropolitan area in summer during which electricity supply usually became short. During Onagawa construction, anti-nuclear movements were relatively strong in Japan. However, the welcoming attitude of the local community paved the path for the smooth construction of nuclear power plants in Japan (Fujigaki 2015). One reason for this positive attitude may have been an expected economic benefit on the local community (Sato 1991). Although Three Mile Island nuclear power plant accident occurred in 1979 had increased the concerns for safety of NPP, the construction had been completed in 1983 and became commercially operational in 1984.

Onagawa NPP is located on the south end of the Sanriku Coast where the local government took a positive decision concerning the siting of the plant due to the expected economic benefit mentioned earlier. According to the historical records, Sanriku Coast has been hit by huge tsunami five times with huge numbers of casualties. The oldest ever recorded is Jogan tsunami, which occurred in AD 869. A more recent one, Keicho tsunami, took place on December 2, 1611 with an epicenter off the Sanriku Coast in Iwate Prefecture. In 1896, the area was hit by the Sanriku earthquake, which reached a 8.5 magnitude. It occurred at approximately 166 km (103 mi) off the coast of Iwate Prefecture. During the twentieth century, the Sanriku Coast was repeatedly struck by two events of tsunami. One is the Showa-Sanriku tsunami in 1933, and the other is the Chili tsunami in 1960.

In the case of the Jogan tsunami, the tsunami has been estimated to flood at least 4 km inland. When Onagawa NPP was designed, these historical disasters had been taken into consideration quite thoroughly in order to determine the ground level of the plant (Sato 2011). A detailed literature survey had been performed to confirm the scale of ancient tsunami. The contribution of one of the distinguished engineers in this process should be mentioned here. His name is Yanosuke Hirai and he was a member of the advisory board for the construction of Onagawa NPP established in 1968. He insisted that the ground level of the plant should be as high as 14.8 m from sea level, while the required level by regulation was about 3 m. He had a strong belief that nuclear power plants had to be prepared against tsunami as large as the Jogan tsunami (Wakabayashi 2016). He knew from old tradition that ancient tsunami came over inland and even reached the shrine near his house in the Sendai plain (Read 2012). When the second unit in Onagawa NPP was planned and designed around 1987, an additional archaeological survey was performed to investigate the history of tsunami leading to a decision to reinforce the slope of seawalls. There is discussion whether this good practice should be attributed to the superior corporate culture of Tohoku Electric Power Company emphasizing the priority on safety or to the existence of one distinguished engineer with excellent foresight. We believe the fact that the final decision by the company to build the plant on the ground as high as 15 m and maintaining their concern of a possible tsunami as high as the Jogan tsunami strongly implies the existence of superior corporate culture pursuing a higher level of safety.

How exactly did Onagawa survive the 2011 earthquake and tsunami? The following is a brief account of how the events unfolded at the nuclear power station. When the earthquake hit the site at 14:46 on March 11, 2011, Units 1 and 3 were at full-powered operation, while Unit 2 was at the beginning of the start-up operation procedure after maintenance. All units were automatically shut down by a safety mechanism that was activated by the "Excessive vertical acceleration" signal. Outside the plant, four out of five off-site power lines were lost due to massive impact of the earthquake. In response to this situation, emergency diesel generators started successfully as designed to provide power to continue emergency operations. Unit 2 reached the cold shutdown state at 14:49 and the other two units continued core cooling by the Reactor Core Isolation Cooling (RCIC) system. When the first tsunami hit the site at 15:29, all units were already sub-critical. The seawater pumps, which play an important role as the final heat sink, were not damaged and the high- and low-pressure water injection functions were preserved. On March 12, Unit 1 reached cold shutdown at 0:58 on and Unit 2 at 1:17. As a result, there was no severe damage in the reactor core and no radioactivity was released from the site.

How did Onagawa manage to avoid Station Blackout (SBO) and fully cool down all its reactors successfully unlike what happened at Fukushima Daiichi? The situation could have been more critical, if not for the decisions that led to positive contributions to mitigate the impact. The most significant pre-condition that prevented the site from reaching a more disastrous situation was the determination of the site level from the sea level. Onagawa was designed to cope with a tsunami height of O.P. + 14.8 m, while the estimated height of tsunami was O.P. + 3 m when Unit 1 was designed (Sasagawa and Hirata 2012). This indicates that the site level was designed with sufficient margin, even though tsunami height records may not contain enough historical information. This should not be simply dismissed as a fortunate factor, but considered as a good practice to make decisions with large uncertainties under conditions concerning risk. Even if the final severity of the damage and the consequences was significantly mitigated by decisions that led to positive contributions with fortunate factors, it is difficult to ascertain the reduction of the severity. In addition, one of the five off-site power lines remained available by chance, a basic pre-condition for processes leading to successful cold shutdowns.

While the site level played a crucial role in saving Onagawa from a catastrophe, a set of good practices institutionalized by the operator organization should be highlighted in revealing the lessons to learn. From the viewpoint both of resilience engineering and sociotechnical resilience, these good practices are described and discussed in the following section.

Averting Meltdown

All of the good practices at the Onagawa nuclear power station that contributed to the successful safety procedures during the 3/11 disaster have been summarized (IAEA 2012) based on interviews and official records of the Great East Japanese earthquake and tsunami that occurred on March 11, 2011. Although it is easy to point out actual events and their consequences, cause–consequence relationships were difficult to determine because of cognitive biases and lost memories over time. Furthermore, it was difficult to determine the root causes of failure and good practices.

With all the information obtained after the disaster, there is a tendency to interpret the accident scenario based on hindsight. To evaluate good practices so that lessons applicable to general situations are learned, the pre-conditions enabling good practices should be elucidated. In addition, factors based on fortune that led to good practices should also be identified to separate good decisions from luck. The pre-conditions and fortunate factors for each good practice are described and discussed in the sections that follow.

Quick Evacuation from Controlled Areas

After the quake, all workers had to evacuate from the controlled area. Checking out from the controlled area requires a surface scan of radioactivity by passing through gate monitors. The gate monitors at Unit 3 became unavailable at that time. The administrator for radiation management took the initiative to bypass the gate monitors and performed manual scanning using a survey meter with the help of a volunteer radiation management administrator from an associated company. The checkout process of seventy workers was successfully completed around 15:30. This is the case when the "responding" played an important role based on the principles of resilience engineering. Although emergency evacuation training drills from controlled areas were practiced frequently, the associated workers were organized and adapted to the situation.

There are pre-conditions that enabled this good practice. The emergency evacuation drills were practiced frequently and experienced volunteer radiation managers were available. The responding actions using manual scanning devices were possible based on the experiences obtained through the emergency evaluation drills. The voluntary and corroborative actions of the radiation experts from an associated company played an important role to realize this good practice.

Alongside the pre-conditions, fortunate factors also supported this effective emergency response. One factor is that the number of workers in the controlled area was not too large. It was partly because of luck that the number of workers in the controlled area was not too large to be handled in the short time before the tsunami came. The other is that the workers unusually remained calm and did not panic as happened in other places during critical situations. If the workers had become panic, well-ordered evaluation would not have been performed and the situation could have become more serious and critical. Although it had not been purely fortunate factors, their calm behavior during emergency was partly dependent on the uncontrolled factor.

Maintaining Communication Channel Through PHS

Personal handy phone system (PHS) terminals were provided to the workers of associated companies. This enabled quick notification of the tsunami warning to the workers in the controlled area and at the seaside. The PHS played a crucial role in maintaining communication after the paging system was rendered unusable by the tsunami. Although this is not directly related to the four main capabilities of resilience engineering, it reinforces the importance of advanced preparation of necessary resources. On one hand, the pre-condition that enabled this good practice is that functionality of PHS system was maintained against the partial loss of power (JANSI 2013). On the other hand, the fortunate factor contributing to this positive outcome is that communication engineers of the associated company were at the site to maintain communication capabilities. From the viewpoint of sociotechnical resilience, this function of PHS played an important role for the informational relations. Although the PHS had not mainly been for use during disastrous situations, it supported the organizational activities by providing information channels within a sociotechnical organization.

Transfer of Authority to On-Site Personnel

The appropriate transfer of authority to the plant site with minimal intervention from the headquarters is considered a good practice in an emergency. Although prompt transmission of required information from the Onagawa site to the headquarters in Sendai was important, the emergency headquarters also should not disturb the on-site emergency procedure. For example, a TV conference system was available as a form of communication between the emergency headquarters and the site. Personnel in the headquarters were aware of the negative impact of using the TV conference, as it may occupy personnel resources, delay emergency procedures, and may degrade the process of taking countermeasures. Based on experience, it was decided that TV conferences should be utilized after initial countermeasures are performed and the on-site situation has come under control.

This good practice was based on shared knowledge among personnel in the nuclear power division learned through experiences at the Onagawa or

Higashidori NPP—the latter is another NPP owned by Tohoku Electric Power Company. The background of this good practice should be mentioned here. As the Tohoku Electric Power Company is relatively small compared to TEPCO, most of the people in the nuclear-related division knows each other personally. This is the reason why mutual trust between the headquarters and the site had meaningfully developed, which resulted in the successful transfer of authority. The people with higher authority in the headquarters in Sendai were certain that the people struggling in Onagawa were capable of managing the situations. In addition to the learning capability, the capability of the organization related to "responding" was the ability to manage communication flow while minimizing interruptions from the headquarters. Again, the preconditions that led to this practice is that the emergency was manageable without support from the headquarters, whereas the fortunate factor supporting this good practice is the fact that the chief engineer of reactor was on duty for the start-up process of Unit 2. This story implies the higher level of resilience in terms of sociomaterial structures as addressed in Amir and Kant. Although the use of the TV conference system had already been proceduralized, they flexibly behaved in relation to its sociomaterial context.

Mitigation Against Basement Flooding of the Attached Reactor Building

When tsunami hit the coast of the Onagawa site, seawater penetrated through the seawater level transmitter boxes that were installed without considering the effect of tsunamis (IAEA 2012). As a result, the reactor building auxiliary area basement was flooded. As the flooding of seawater would have rendered important pumps unusable, the remedial action was to drain seawater and protect the pumps. This began around 17:00 on March 11 with the help of workers from associated companies. Pumping seawater from the third basement floor to the ground was a difficult task. The water had to be pumped up 23 meters and the lifting height of a single pump was insufficient. An intermediate tank was temporarily installed and two pumps were used in series. The balance of the flow rate of the first pump and the second pump was accomplished through trial and error.

To protect the Reactor Building Closed Water System (RCW) pump, which is extremely important to maintain various safety functions, workers sandbagged the flooded water. The frontline personnel were instructed to fill 100 sandbags and place them around the pump. The situation changed frequently, which required the workers to adapt quickly. Initially, they had to change the arrangement of sandbags to provide additional protection for the air conditioning unit of the pump. When seawater began to flow into the protected area from another drain piping, additional sandbags were needed to further protect the area. When another instruction to fill an additional 100 sandbags was issued, they had already filled them based on proactive decision-making. This prevented further flooding and protected the pump.

In retrospect, this is a typical example of good practices based on combined capability related to sociotechnical resilience. The experienced workers cooperated to perform the difficult task of draining seawater and prevented damage to the pump. The workers had to make decisions according to the dynamically changing situation and anticipated a change for the worse. They proactively prepared the required resources (sandbags) in advance, showing the capability of "anticipating" and "responding." The pre-conditions for this good practice should be noted here as well. An important factor is the cooperation of the skilled workers of the associate company. This successful cooperation with associate companies was realized based on the already developed good relationship of Tohoku Electric Power Company and associated companies. Although their relationships were of contractor and subcontractor, they definitely had some kind of culture and cooperative mind in common. In addition to this are two fortunate factors. One is the speed of flooding, which was not too fast and the remedial action worked in time. The other is that the number of workers available at the site was exceptionally large because of various ongoing construction work, such as a construction of a new anti-seismic office building.

Accepting Locals into the Site

Another contributing factor contributing to saving Onagawa from falling into worse situations is the operator's decision to let the locals enter the facility after the tsunami struck the area. Locals affected by the earthquake and tsunami came to the site for assistance in the evening of March 11. The superintendent of the Onagawa NPP immediately decided to accept them into the site and provided them with food, shelter, and necessities. This decision was not taken lightly, with existing security requirements against terrorism as directed by relevant government authorities. They accepted about seventy people on the first day. Evacuees were initially housed in a former office building and subsequently transferred to the gymnasium due to security reasons. Increasing number of evacuees from the local area near the site came for help and a total of 364 people, about 40% of the local community, stayed at the site on March 14.

The decision to accept evacuees into the site was made by the superintendent of the NPP. This was not an easy decision because a process of security check could not be performed strictly and it required additional various resources. The following process of taking care of evacuees was well organized, which could be related to the capability of "responding." Nevertheless, accepting locals to the site provided an unexpected opportunity to reopen damaged roads to the site. The three traffic routes from Onagawa to the city of Ishinomaki were impassable after the earthquake and the tsunami struck. There was an urgent need to reopen the roads to transport essential resources. Houses and ships that were carried by the tsunami had blocked access. Although there were available heavy machineries for road work at Onagawa, the task of clearing roads required permission from the owners of the houses and ships. It was fortunate that most of the owners were evacuees at the site and the permissions were easily obtained. The road clearing work started on March 13 and completed on March 15. Again, there were pre-conditions that enabled this good practice. First, the decision to accept evacuees into the site was later confirmed by relevant government authorities. Second, the operator had established a good relationship with the local community, which created a trust between the locals and the operator. Further contributing to this situation was the fact that the possibility of the release of radioactive material was not high. Also considered a lucky factor is that none of the evacuees allowed to enter the facility had ill intention, despite a lack of security check. Finally, it was widely considered by personnel in the Tohoku Electric Company that it was very fortunate that the earthquake happened on a Friday afternoon. Most of the workers were on their duties and able to provide maximum work force for remedial actions.

Conclusions

In this chapter, we have described a fascinating story of sociotechnical resilience by focusing on the successful attempt of the Onagawa nuclear power plant to avoid a massive scale of nuclear meltdown. Onagawa is a remarkable case study because despite the impact of the 2011 Great East Japan earthquake and tsunami on the plant, it was able to mitigate the impact and dodge the unfortunate situation that led Fukushima Daiichi to nuclear disaster. Using the viewpoint of resilience, this chapter has argued that the ways in which the Onagawa nuclear power station managed to reduce the severity of impact caused by the natural disaster were determined by a combination of pre-conditions and fortunate factors. We highlighted both factors, and emphasize the pre-conditions as inherent characteristics of sociotechnical resilience. We combined two theoretical frameworks to explain the resilience characteristics of Onagawa. One is resilience engineering, the other is sociotechnical resilience. As noted, while it is easy to point out flaws in actions taken after the accident, this chapter identified positive contributions to mitigate accidents as a lesson to prevent future disaster. The good practices were described with their underlying pre-conditions and fortunate factors. We concluded by emphasizing decision-making as an important element contributing to reduction of damage in severe situations.

Notes

1. The 3/11 disaster is shorthand for the 2011 Great Earthquake, tsunami, and nuclear disasters that hit Japan on 11 March 2011.
2. For detailed information, please see the website of Tohoku Electric Power Co. Inc. at www.tohoku-epco.co.jp/electr/genshi/npi/onag-e.htm, accessed December 20, 2017.

References

Amir, S. & Kant, V. (2018). Sociotechnical Resilience: A Preliminary Concept. Risk Analysis, 38: 8–16. https://doi.org/10.1111/risa.12816.

Fujigaki, Y. (2015). The processes through which nuclear power plants are embedded in political, economic, and social contexts in Japan. In *Lessons From Fukushima* edited by Y. Fujigaki. (pp. 7–25). Springer International Publishing.

Hatamura Y. (Ed.) (2012). Final Report of Investigation Committee on the Accident at Fukushima Nuclear Power Station of Tokyo Electric Power Company, July 23.

Hollnagel, E. (2014). Safety-Iand Safety-II, The Past and Future of Safety Management. Ashgate Publishing.

Hollnagel, E. (2017). Safety-II in practice: developing the resilience potentials. Taylor & Francis.

Hollnagel, E., Woods, D. D., & Leveson, N. (2006). Resilience engineering: Concepts and precepts. Ashgate Publishing, Ltd.

Hollnagel E., Paries J., Woods D. D., & Wreathall J. (Eds.) (2011). Resilience Engineering in Practice, Ashgate Publishing Co.

IAEA Mission Report (2012). IAEA Mission to Onagawa Nuclear Power Station to Examine the Performance of Systems, Structures and Components Following the Great East Japanese Earthquake and Tsunami. IAEA Department of Nuclear Safety and Security Department of Nuclear Energy.

ISO/IEC (2014). Safety aspects—Guidelines for their inclusion in standards, GUIDE51. Available at http://isotc.iso.org/livelink/livelink/open/8389248, accessed 10 December 2017.

JANSI (2013). "Report on the responsive actions of Onagawa NPP and Tokai Daini NPP against the Great East Japan Earthquake and tsumani. Aug, 2013 (In Japanese). Available at http://www.genanshin.jp/archive/disastersitereaction/data/OnTk_Hokoku.pdf, accessed 10 January 2018.

Kitamura, M. (2014). Resilience engineering for safety of nuclear power plant with accountability. Resilience engineering in practice, 2, 47–62.

Kitamura, M. (2016). Supplementary Remarks on Lessons Learned from the Great East Japan Earthquake, IFAC-PapersOnLine 49–19, 257–260.

Okamoto, T. (2017). Reorganization, Deregulation, and Liberalization: Postwar Development of the Japanese Electric Power Industry and Its Change after 11 March 2011. *Technology and Culture*, *58*(1), 182–193.

Read, R. (2012). "How tenacity, a wall saved a Japanese nuclear plant from melt-down after tsunami." Available at http://www.oregonlive.com/opinion/index.ssf/2012/08/how_tenacity_a_wall_saved_a_ja.html, accessed 15 December 2017.

Sasagawa, T. & Hirata, K. (2012) Tsunami evaluation and countermeasures at Onagawa Nuclear Power Plant. In *Proc. 15th World Conf. Earthquake Engineering*, Lisbon, Portugal, 24–28 September 2012.

Sato, H. (1991). 女川町史(続編)[*History of Onagawa* (Sequel)]. 女川町史続編編さん委員会. [Editorial Committee of History of Onagawa, Town of Onagawa]

Sato, S. (2011). Social Responsibilities of Engineers Engaged in Mega Technologies -Impressions received through Communications with the Media about the Great East Japan Earthquake. Available at http://www.iee.jp/wp-content/uploads/honbu/39-doc/201209-dayori2.pdf, accessed 20 December 2017.

Takahashi, M. and Kitamura, M. (2013) Actions Contributed to Disaster Level Reduction of the Fukushima Accident. Paper presented at the 5th Resilience Engineering International Symposium, Soesterberg, The Netherlands, 25–27 June.

Wakabayashi, T. (2016), Good practice at Onagawa during 3.11 East Japan Great Earthquake, Maintenology, 15(3):14–19.

Yoshizawa, A., Oba, K., & Kitamura, M. (2015). Experiences in Fukushima Daiichi Nuclear Power Plant in Light of Resilience Engineering. In *Proceedings of the 6th Resilience Engineering Symposium*, Lisbon, 22–25 June 2015.

Yoshizawa, A., Oba, K., & Kitamura, M. (2016). Lessons Learned from Good Practices During the Accident at Fukushima Daiichi Nuclear Power Station in Light of Resilience Engineering. *IFAC-PapersOnLine*, *49*(19), 245–250.

PART V

Urban Life

CHAPTER 11

An SME-Driven Approach to Adopting Measures of Flood Resilience: A UK-Based Perspective

Bingunath Ingirige and Gayan Wedawatta

Introduction

Small and Medium-sized Enterprises (SMEs) are abundant in many sectors in the United Kingdom. According to the Department for Business, Innovation and Skills (BIS 2015), SMEs account for at least 99% of the businesses in most of the industry sectors. Although SMEs are widespread in many industry sectors, their vulnerability to many challenges remains very high. For instance, research has found that SMEs suffer the most in times of crisis and are the least prepared of all organisations (Ingirige et al. 2008). SMEs are severely affected by external disruptions and without a coherent strategy it is difficult for policy makers to address the problem. For example, a recent survey found the financial cost of severe weather events to have been just under £7000 for each affected SME (Federation

B. Ingirige (✉)
University of Huddersfield, Huddersfield, UK
e-mail: b.ingirige@hud.ac.uk

G. Wedawatta
Aston University, Birmingham, UK

S. Amir (ed.), *The Sociotechnical Constitution of Resilience*,
https://doi.org/10.1007/978-981-10-8509-3_11

of Small Businesses 2015). Climate change is set to make things worse for the SMEs in that according to the Intergovernmental Panel on Climate Change (IPCC) (2007) there seems to be more widespread extreme weather events (EWEs).

Policy making in the United Kingdom (see Defra 2011) encourages at-risk communities to learn and live with and adapt to flooding rather than over-depending on government investments on expensive flood defense schemes and over-relying on insurance. Measures of protection from flooding could take the form of structural measures or non-structural measures. Structural measures involve making adjustments to the property and non-structural measures are the processes and considerations on continuity of the business and coping with the flood risk. Although theoretically adopting such measures seems quite rational, there have not been any significant studies undertaken on property-level measures as a whole and in particular the effect of non-structural measures of flood protection, which tend to be the more softer measures that can be progressively implemented considering the unique context of small businesses. The aim of this chapter therefore is to generate some theoretical arguments and a synthesis of practical cases on the role of the overall property-level measures and specifically the non-structural soft measures of resilience as a natural and a cost-effective way of risk response for small businesses at risk of flooding. The chapter lays emphasis on how such changes can be incorporated at a policy-making level to generate a coordinated and a cohesive approach to improving resilience in small businesses.

SMEs and Sociotechnical Resilience

Literature on SMEs reveals that a universally agreed definition of an SME is still to be arrived at (BERR 2007; Gunningham 2002; Lauder et al. 1994). There are various operational and theoretical/economic definitions, based on either quantitative, qualitative or a mix of quantitative and qualitative criteria. Whilst the United Kingdom does not have an official definition for an SME (Ayyagari et al. 2003), several operational and theoretical definitions exist. For statistical purposes, the BIS defines an SME as "any enterprise with less than 250 employees" (BIS 2010). Section 382 and section 465 of the Companies Act 2006 define a small business and a medium business for the purpose of accounting requirements. Accordingly, an SME can be identified as an organisation that employs not more than 250 employees and/or has an annual turnover of not more than £22.8

million and/or has a balance sheet total of not more than £11.4 million (Companies Act 2006). The current EU definition of an SME is "an enterprise which employs fewer than 250 persons and which has an annual turnover not exceeding €50 million, and/or an annual balance-sheet total not exceeding €43 million" (European Commission 2006). The Organisation for Economic Co-operation and Development (OECD) defines SMEs as "non-subsidiary, independent firms which employ fewer than a given number of employees" (OECD 2000). Whilst this upper limit may vary across countries, industry sectors and so on, the upper limit is commonly recognised as 250 employees in the United Kingdom. Accordingly, an SME can be generally identified as one that employs less than 250 employees.

These SMEs represent a significant portion of the UK economy as they amount to more than 99% of private sector enterprises. Whilst their individual contribution to the economy may be insignificant due to their smaller scale of operations compared to large businesses, their cumulative impact is highly significant (International Institute for Sustainable Development 2004). Their contribution is important not only in terms of numbers, but also in terms of turnover (47%) and employment generated (60%) annually (BIS 2015). SMEs are thus considered as the backbone of the UK economy (Crichton 2006; Lukacks 2005) due to their significant economic importance. Importance of SMEs is also recognised in terms of fostering competitiveness (European Commission 2006), technological progress and innovation (Tilley and Tonge 2003), and more equal distribution of income and wealth (Hallberg 2000) among a multitude of other factors.

According to Robbins et al. (2000) SMEs are important to the economic vitality of cities, states and countries due to their significant number and employees. For instance, the United Nations Development Programme (UNDP 2013) highlighted the crucial role played by SMEs in relation to disaster/climate impacts including:

- A strong SME sector promotes a country's resilience to shocks by broadening and diversifying the domestic economy
- Resilience of livelihoods is intimately related to the resilience of SMEs and their ability to promote a healthy local economy after disasters, making their recovery crucial for broader economic recovery
- Role in mitigating the impact of disasters in the broader community and encouraging its recovery

A flood-affected SME can experience a multitude of impacts varying from damaged property/stocks and loss of income for an extended period to business closure/failure (Ingirige and Wedawatta 2014; Marks and Thomalla 2017). Developing resilience is important if such impacts are to be minimised. Further, SMEs are often involved in complex supply chains of larger businesses. In the construction sector, for example, hundreds of SMEs come together to deliver large construction projects. In manufacturing sectors, SMEs can be responsible for delivering key components of a final product. As such, flood impacts on a SME like that can create a wider impact on its supply chain, an impact much larger than that on the individual SME itself.

Resilience of the SME sector therefore is an important issue for regions/ countries vulnerable to flooding. Whilst even the larger businesses are affected and in fact are required to be resilient, SMEs warrant special attention due to their inherent characteristics. Blackburn and Smallbone (2008) argue that SMEs cannot be treated as scaled-down versions of large businesses as they have unique characteristics that determine how they respond to challenges and opportunities; like that of flood risk. Following Amir and Kant (2018) in characterising sociotechnical resilience, SMEs embody a set of features in which its sociomaterial structure is inherently flexible when responding to environmental hazards. These advantageous features are due to the size and strong social networks that allow SME organizations to adapt to the changing environment.

Vulnerability of Small Businesses to Flooding

From the preceding it is clear that it's a "no-brainer" that the SME existence and their continuous operations are to be sustained so that they continue to benefit local communities and economies. However, despite a community's motivation to keep the clock ticking in SME operations, it is widely accepted that they tend to display vulnerability in facing up to various conditions prevailing in a country's economy resulting in sudden failures. Ability of SMEs to turn around their companies is constrained due to their limited access to financial resources and capital (Kirchhoff 1994). Due to the inherent qualities pertaining to their operations, the sudden failures in SMEs could also result due to disruptions caused by EWEs such as flooding. A recent Federation of Small Business study (FSB 2015) into "a more resilient small business community" found that two thirds of the surveyed small businesses were impacted by severe weather during the last three years and about 93% of them believe that severe weather poses a risk to some part of their business. The winter 2012 floods in the United Kingdom caused economic damage to the extent of £620 million, where it has been assessed that the cost to businesses stood at £200 million, with £33 million in losses related to economic costs, such as lost working days (Climate ready 2014). It is also known that a large number of SMEs either do not take insurance or are under-insured against weather-related risks, thus hampering their abilities to recover from sudden failures and disruptions.

SMEs' ability to effectively react to various EWEs are often affected due to lack of planning and preparedness (Bhattacharya-Mis and Lamond 2014); vulnerability to cash flow interruptions; lack of capital for recovery; ineffectual interactions with national agencies; infrastructure problems (Runyan 2006); individual attitudes and organisational culture (Petts et al. 1998); access to expertise; business sector; and perceived exposure to risk (Yoshida and Deyle 2005). Collectively, these factors determine the adaptive capacities and the overall behaviour of SMEs. For SMEs the financial losses can result in total devastation, not only of their businesses but also the immediate communities and small townships that are dependent upon them. Therefore, considering the vulnerability of SMEs can help in identifying measures that can enable them to cope with extreme events, such as flooding, better.

SMEs in general are said to be more vulnerable to the effects of flooding than larger businesses: as although being unable to trade for some time might not have a major impact on the latter, but it could have a significant effect on an SME reliant upon a short-term cash flow. Factors like lack of resources and the small scale of operations are likely to restrict their ability to respond to a flood situation quickly and effectively. Further, as a majority of SMEs tend to be rooted in local communities (Bannock 2005) their owners are likely to be affected as residents as well as business owners (Runyan 2006). This puts extra burden on SME owners, having to deal with flood impacts on two fronts. Adverse effects of flooding on SMEs can be substantial and can even put them out of business. For example, a survey of businesses operating in the Cumbria region found that a majority of businesses that have ceased business following flooding and related EWEs in 2009 and 2010 were found to be SMEs (Wiseman and Parry 2011).

These vulnerabilities are further escalated by the fact that SMEs in general are less prepared to face the eventualities of flooding and related impacts. Crichton (2006) found that 70% of small businesses in high-risk areas were not concerned that flooding might affect them. Further, a similar percentage of businesses were found to have no form of business continuity plan in place in the event they are flooded. These two figures summarise the level of concern of UK businesses on flood risk and adaptation. It has to be noted that this is not limited to the UK context alone, but has been observed in other contexts; for instance, Germany (Kreibich et al. 2007, 2008), France (Pivot and Martin 2002), Australia (Gissing et al. 2005), and the United States (Tierney 1995) as well; where small businesses were found to be less concerned about flood risk adaptation. Those with previous flood experience are more likely to be prepared for such events, at least in some form, than those who have not been affected previously. This is because of the fact that some SMEs implement various adaptation measures after being affected by a flood event.

Property-Level Resilience Measures and the Importance of Softer Measures of Resilience

First line of defense in the case of flooding comes from community-level flood management strategies that provide protection to a community. These schemes, however, may not be available in all places at risk of flooding, and some level of residual risk may remain even at places where there are such schemes (Environment Agency 2009b; Lamond et al. 2017). As such, property owners, including SMEs, are increasingly urged to protect their properties by implementing property-level measures. Property-level flood protection measures can be either resistant or resilient (Bowker et al. 2007). Resistant measures attempt to prevent flood waters entering the property, whereas resilient measures attempt to minimise the impact of flood waters on property (Bowker et al. 2007). Collectively, these can be identified as "structural" or "hard" flood protection measures for SME property owners. As far as businesses are concerned, one of the recommendations of the Pitt Review (2008) was to promote business continuity by encouraging the uptake of property-level flood resistance and resilience measures by businesses. Further, the Pitt Review (Pitt 2008) recommended building regulations to be revised to ensure new and refurbished buildings in high flood risk areas are flood resistant or resilient. Although availability of information, guidance and standards on property-level protection measures against flooding for existing as well as new buildings have seen an influx over the years, implementation of such strategies still seems to remain quite low. This can be attributed to presence of various barriers that hinder such implementation by businesses.

Complementing the structural measures, non-structural measures seek to "manage risk by building the capacity of people to cope with flooding in their environments" (Jha et al. 2012). For businesses, including the SMEs, these are their business continuity strategies. According to Crichton (2006), businesses are likely to implement various generic coping strategies that aid business continuity, rather than property-level adaptation strategies against flooding. This is in line with what Berkhout et al. (2004) discussed in relation to climate change, where instances of adaptation to climate change were noted in businesses, which have been implemented due to commercial purposes rather than having climate change in mind. However, as generic strategies for business continuity can only limit adverse consequences on a flood hit business and aid recovery process; rather than preventing/limiting damage to property and its contents, some form of property-level protection is desirable if a business is located in a high flood risk area. A report on businesses in Cumbria (BMG Research 2011) affected by 2009 flooding found that more than half of the businesses (52%) that moved to temporary premises—as their premises were flooded—have not returned to the original premises even after six months from the event. This alone suggests how long it can take for a flooded property to be reinstated and, thus, how important it is to have property-level protection measures in place, in order to minimise damages and aid quick recovery. Ingirige and Wedawatta (2014) presented the balance between structural (property-level) and non-structural (Business continuity) measures that SMEs can implement in order to achieve a required level of flood resilience. Figure 11.1 demonstrates how an SME could transit between adopting structural and non-structural measures to adapt to flood risk. Adaptation of business organisations to flood risk, especially SMEs who often are resource constrained, is not always straightforward (Wedawatta and Ingirige 2012). A multitude of factors are likely to influence the processes of decision making and implementing, both as barriers and drivers in the case of individual SMEs (Wedawatta et al. 2011).

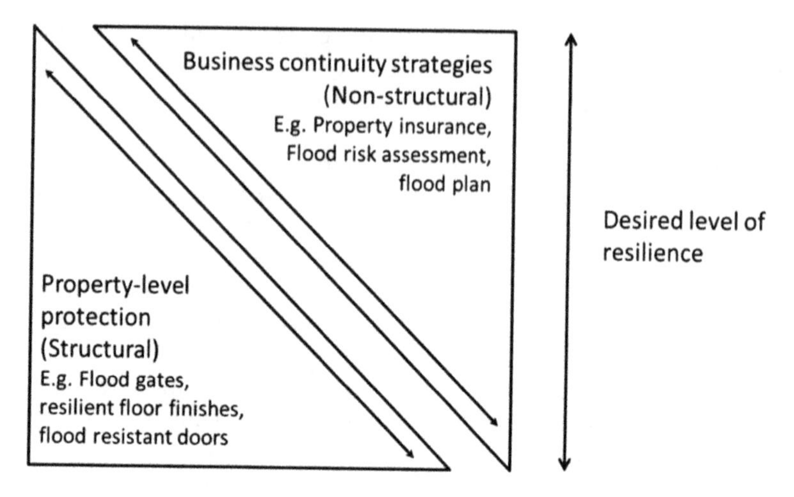

Fig. 11.1 Achieving a desired level of flood resilience via structural and non-structural measures by SMEs. (Source: Ingirige and Wedawatta 2014)

UK Policy-Making Context and Theoretical Arguments

The increased frequency of flooding, the growing number of properties being constructed on floodplains and the increase in urban dwellings suggest that these statistics are set to worsen in the future. This is further confirmed by the growing scientific consensus that climate change is expected to amplify the prevalence and severity of flood risk, due to changes in winter precipitation, rising sea levels, storm surges and other extreme weather events (Evans et al. 2004; IPCC 2007; Stern 2007). In the United Kingdom, the effect and impact of flooding is highly topical and the Strategic Defence and Security Review conducted in 2010 (HM Government 2010) found that flooding and coastal erosion is an item at the top of the agenda having a significantly high risk in terms of economic, social and environmental consequences. Therefore, adopting measures to avoid or to control flood disasters and disruption has received much attention from policy makers and scientists.

From a policy-making angle, there is also a shift in exploring how to sustain life and ordinary business amidst the changing climate concentrating more at an individual property level (Defra 2008, 2011; Environment Agency 2009a), thereby empowering communities rather than focusing too much attention on large-scale flood defences. Defra's (2011) report identifies that communities at risk of flooding should learn to live and adapt to flooding by implementing adjustments to their property (structural measures) and processes (non-structural measures such as business continuity strategies) rather than relying totally on insurance or the government to invest in expensive schemes. In this context, some researchers argue that learning to live with rivers (ICE 2001), living with risk (UNISDR 2004) and knowing risk (UNISDR 2005) are becoming increasingly accepted, adopted and implemented. Although theoretically this shift seems quite rational, there have not been any significant action plans or initiatives that have emerged to implement the measures in practice. It seems that the current policy-making initiatives that target certain behavioural change among small businesses falls short of practical mechanisms, which are sympathetic to the needs of SMEs. The Defra's Flood and Coastal Erosion Risk Management (FCERM) strategy (Defra 2011) on empowering communities emphasises that local communities should:

(1) take part in any public consultations (e.g., on any future local strategies) and otherwise working closely with local authorities to ensure local views and ideas inform their decisions; and
(2) make sure they are represented in local flood risk management partnerships (or equivalent) and community resilience initiatives, such as flood or coastal action groups, preparing community flood action plans, or promoting schemes to make properties more resilient to flooding, or helping the community adapt to coastal erosion (Defra 2011, p. 14).

The preceding actions within the FCERM strategy are intended to empower communities in decision making to mitigate/adapt taking into consideration of the flood risk. The stakeholder engagement and participatory processes built around implementing the FCERM strategy could be further customised to ensure that the measures are sensitive and appropriately aligned with the case of SMEs. The SME businesses could then develop an understanding of the sensitivity of the property-level adaptation strategies and business continuity strategies towards the overall profitability and sustainability of the enterprise. Given that practically it is very difficult to protect every property at risk of flooding through community-level flood defence schemes, adapting individual properties at risk of flooding by implementing property-level measures such as installation of flood barriers (door guards, airbrick covers, etc.) and putting up sand bags to resist the effect of flooding, or by installing resilience measures such as installing concrete floors as opposed to carpets or timber floors so that the property owners can return quickly to the property once flood waters recede, can be very important (Environment Agency 2009a).

Practical Cases of Implementing Resilience Measures and Synthesis

Cockermouth Flood Resilience/Resistance Measures

Cockermouth in Cumbria has experienced a series of flood events in recent times. A study was conducted on the impacts experienced by SMEs and their recovery following the 2009 flood event in Cockermouth (Wedawatta et al. 2012). The flood event in November 2009 affected about 700 residential properties and 225 businesses directly (Cumbria County Council 2010; Tickner 2011), and many others were affected indirectly. According to the Cumbria Intelligence Observatory (2010), 80% of the businesses in Cockermouth were affected by flooding. Several other areas in the Cumbria region were also severely affected by the 2009 flood event, but Cockermouth was said to have suffered the worst impacts with flood depths in excess of 1.5 m reported (Environment Agency 2009a).

A survey including 48 SMEs and case studies of flood-affected SMEs were conducted to study the impacts of flooding, post-flood reinstatement and recovery of affected SMEs (Wedawatta et al. 2012). Although the study found that many SMEs have not utilised the opportunity presented by the flood damage to reinstate their properties integrating property-level adaptation measures, thus enhancing their level of resilience to a similar flood event in the future, instances of some or extensive adaptation were also noted. In general, the level of uptake of "structural" measures was limited to a handful of SMEs surveyed. Whilst SMEs are generally adept in making business decisions, they are in a better position to assess non-structural or soft measures as opposed to structural measures, which require specialist input from built environment professionals. Case studies of SMEs revealed that some businesses whose premises were significantly damaged by flooding, and thus required extensive reinstatement, have not received adequate guidance on flood protection thus hindering their ability to implement structural measures. The role of the built environment professionals in aiding the uptake of structural measures by SMEs is highlighted by Lamond et al. as well (2017). Further, the understandable desire of the affected SMEs to reinstate their properties quickly and restart their business activities has resulted in them being hesitant to consider new structural protection measures to their properties, which require additional time to be planned (Wedawatta et al. 2012).

In terms of non-structural measures, SMEs often rely heavily on insurance. One of the key findings to emerge from the Cockermouth study was the inability to rely solely on insurance in the future. SMEs affected by flooding have seen their insurance premiums and excesses increase considerably (Wedawatta et al. 2012). The impact on insurance excess, the cost that the policy holder has to bear when a claim is made, was more significant as this will reduce the protection available in a future flood event. Excesses increasing from £0 to £5000, £100 to £10,000, and £1000 to £15,000 were quoted, although, in a few cases, the increase was in the range of 20–25%. High increases in insurance excesses are likely to make property insurance less attractive for businesses at risk of flooding, as they limit the damages claimable in future flood events. The issue is further escalated by non-betterment clauses and inactivity of insurers when it comes to property-level flood protection (Douglas et al. 2010), leading to less resilient reinstatement and property-level flood protection. Previous evidence suggests that SMEs rely heavily on insurance to recover from flood damage (Crichton 2006). But the evidence from the Cockermouth study suggests the importance of addressing remaining risks through other resistance and resilience measures, as SMEs can no longer rely solely on insurance. Whilst in some areas, protection may come from community-level flood defences, but at the individual level, it is important that SMEs adapt to the risk by addressing flood risk in their business decision making and implementing property-level flood protection measures. The Cockermouth study also showed the heterogeneous nature of SMEs whereas some SMEs have better prepared their properties and businesses through various property-level and business continuity measures to suit their context and capabilities whereas some of their neighbours have not done so.

Braunton Flood Resilience/Resistance Measures

The short case studies in Braunton, North Devon provided further value in understanding unique measures adopted by SMEs affected by flooding. SMEs in Braunton have experienced several flood events during the last few years. Of particular interest were the two recent episodes of flooding, once in 2009 and again in 2012. Both events occurred in the month of December with flood water reaching a higher level and remaining standing for a longer period in the 2012 event. This event proved catastrophic to the small businesses in Caen Street due to its low-lying locale (the main high street in the Braunton village). Six months before the flood event, the Environment Agency completed an enhanced community flood defence scheme, which the community perceived as a "seal of confidence" that the area would be flood proof (Ingirige and Russell 2015). However, the 2012 winter flood completely overwhelmed the designed standards of the scheme, which was of an indicative standard of resisting a 1 in 100-year flood event. The resultant flood event was not only caused by overflowing of the river but lack of drainage and heavy rainfall combined with high levels of ground saturation during the winter, which contributed towards building up of surface water flooding within the area.

Seven SMEs and their representatives were interviewed, utilising an SME resilience toolkit produced as part of the Engineering and Physical Sciences Research Council (EPSRC)-funded Community Resilience to Extreme Weather (CREW) project in 2011. The interviews were digitally recorded and then transcribed verbatim. These SMEs consisted of a bakery, card shop, restaurants, newspaper shop, estate agent, optician and a sports shop, which are typical SME businesses found in a typical village or township.

As part of the Braunton study there was overwhelming evidence that SMEs are entities having specific needs in improving their flood resilience based on factors such as their typical business activities, their location and their ownership. Their levels of preparedness differed substantially. In the SMEs own words, what they valued was more of a "practical and emotional assistance" to improve themselves to better prepare for future flood events. One of the important findings was the consideration that in certain instances it was important that the whole community should pull together to make a difference. For instance, one of the case study representatives identified that flood protection needed to be a collective responsibility of the SMEs. The following comment identifies the sophistication of the SME representative's understanding of flood resilience

> *The only [way] we can be [less] vulnerable would be if we could make this an island of security, which we can't because not everybody has put flood gates up. My next door neighbour hasn't, he doesn't own the building, he possibly can't afford to do it. For whatever reason he does not have any flood gates, which means when water comes in to him it will soak through to my wall.*

Such a level of understanding results in undertaking "sophisticated resilience measures" compared with some of the entry-level measures proposed by the majority of SMEs. Paton's (2007) grouping of resilience indicators is similar to this. The entry-level measures are the popular choices that were found across six of the seven SMEs studied. For instance, a popular measure adopted by SMEs was keeping sand bags at hand and being prepared to meet any emergencies. This is perceived as an entry-level strategy as this was a standard approach without much consideration of the SME's individual business.

The SME interviews identified several structural measures that they take into account with the specific business context in mind, moving from adopting a standard blanket procedure such as raising electrics, having sand bags at hand and so on, to one that involves careful thinking and some unique measures consistent with the specific business. These can be termed as secondary measures that gradually build up on the standard ones. These tend to be mainly structural but with an element of non-structural soft decision-making or knowledge generation combined within the structural measures. The SME interviews support the adoption of the measures given here.

- Recognising value of stock/making informed decisions on flood products
- Special mechanisms to protect valuable items/stock
- Equipment to speed up salvaging of expensive stock
- Easily removable wall panelling
- Knowing the locations of sandbag emergency supplies

The move from entry-level standard structural measures to adopting a combination of structural and non-structural measures seems a step forward in the sophistication scale for SMEs, particularly in line with the business continuity planning process. If one goes beyond the face value of how they are interpreted or described, they further involve attitudinal considerations and strong mental models. For instance, the easily removable wall panelling might sound like a purely structural measure, but the SME estate agent that used that method used it in conjunction with business continuity needs of the business. The wall panelling protects the expensive plastering of the walls thereby reducing the disruption caused by prolonged repairs to the building after a flood event. Easily removable wall panelling can be removed and replaced in a few hours without disrupting its business operations.

Braunton therefore presented a very good opportunity to document the different practicalities faced by the SME community in facing up to the challenges of flooding. Such practicalities, measures and challenges that SMEs face will help with documenting good practice across other similar contexts. Considering the fact that Braunton is a typical village where there is a lot of potential to transfer good practice and learning to other areas in the United Kingdom.

The Braunton study also adds value as it is in line with the current government policy of the communities being empowered to learn and live with the flooding (as covered in DEFRA 2011) so that small businesses are able to develop their coping mechanisms for preparedness and continuity of their businesses.

The Important Nexus Between Business Context and Softer Measures of Resilience

Initial theoretical insights and case study discussion shows resilience as a complex concept that cannot be adequately assessed in a single measure. Different measures are needed for different systems. According to Multidisciplinary Center for Earthquake Engineering Research (MCEER) (2005), the definition of resilience involves technical, organizational, social and economic dimensions. While technical resilience refers to the performance of physical systems to disasters (McDaniels et al. 2008), the softer measures can be a combination of sociotechnical systems. McDaniels et al. (2008), for instance, show that the combination of the learning effect of a technical system for resilience could make up the sociotechnical system. Therefore, resilience cannot be perfectly delineated to hard and soft as most measures of resilience tend to occur within a sociotechnical context (Amir and Kant 2018). Nevertheless, softer measures within our chapter refer to measures that are predominantly social.

The key message delivered by the case study discussions is that flood mitigation/adaptation measures are not based on a one-size-fits-all strategy. SMEs have unique needs as demanded by the business and the environment that they are in. As per MCEER's (2005) study of resilience components, the business and the environment context addressed within the study tend to align with the organisational and the economic dimensions. Our studies showed that the softer side to resilience is emerging as a prominent feature as opposed to hardware or technical measures (within a sociotechnical perspective). These soft measures of resilience seem more pertinent within government calls for empowerment of the community to undertake flood resilience [as per the FCERM strategy in DEFRA (2011) mentioned earlier]. The softer measures tend to demonstrate a strong link to the understanding of unique SME needs and adopting and implementing seems an outcome of a conscious approach rather than a blanket strategy of risk management. Any future adaptations are also likely to be delivered on a business-driven basis aligned with a business strategy rather than in a standardised risk-management basis.

A risk transfer approach such as insurance tends to be a short-term measure. Our case studies show that SMEs tend to prefer having more control than a mere risk transfer approach. More long-term sustainable measures that are sympathetic to the needs of the SMEs are therefore desired to further buy in and popularise the policy-making messages.

The results also show that measures or strategies adopted in smaller pockets within large communities do not result in any added value. Instead the whole community should pull together and the approach should be a coordinated and a cohesive one so that the real value of the intended strategy is delivered and policy making is established within the context of flood risk management.

Conclusion

Small businesses play a vital role in economic development within a local setting. Such a role at a local level paves the way for a major economic drive within a country. Hence, government policy makers, private sector and trade organisations would like to promote small businesses in an area. At the same time, it is a well-known fact that SMEs display a degree of vulnerability to the effects of hazards, as their ability to naturally cope with them is limited. This chapter therefore looked into improving their current coping capacities through a series of resilience measures and how SMEs perceive them in their practical use given that the business considerations might not harmonise well with the adoption of the measures of resilience and longer-term sustainability of their business and operations may not be one of their key goals considering that they are, after all, SMEs.

After conducting case studies within two vulnerable communities subjected to the risks of flooding and SMEs who have faced recent flood events, it clearly showed that the unique business environments dictate their levels of coping capacity and vulnerability and there was variability in adopting some of the resilience measures from a purely structural or non-structural to a combined structural and softer non-structural measures. From a policy-making angle this was proving to be challenging, as policies could not be standardised and they have to be unique to a business. However, interestingly the case studies showed some of the unique behaviour patterns of sociotechnical resilience among the SME community. A key challenge for individual SMEs in enhancing their sociotechnical resilience is to identify the best solution that suits the context of their business, level of risk and their capacities. Amidst conflicting business priorities and limited resources, right help therefore needs to be provided to individual SMEs to enable them tackle this challenge.

References

Amir, S. & Kant, V. (2018). Sociotechnical Resilience: A Preliminary Concept. Risk Analysis, 38: 8–16. https://doi.org/10.1111/risa.12816.

Ayyagari, M., Beck, T. & Demirgüç-Kunt, A. (2003). Small and Medium Enterprises across the Globe: A New Database. *World Bank Policy Research Working Paper 3127*, 1–34.

Bannock, G. (2005). The Economics and Management of Small Business: An International Perspective. *London*. Taylor & Francis Routledge.

Berkhout, F., Hertin, J. & Arnell, N. (2004). Business and Climate Change: Measuring and Enhancing Adaptive Capacity. *Tyndall Centre Technical Report 11*. Oxford: Tyndall Centre for Climate Change Research.

BERR. (2007). SME statistics for the UK, 2006 – Methodology and accuracy: Technical note. Enterprise Directorate Analytical Unit, Department for Business Enterprise and Regulatory Reform (BERR).

Bhattacharya-Mis, N. & Lamond, J. (2014). An Investigation Of Patterns Of Response And Recovery Among Flood-affected Businesses In The UK: A Case Study In Sheffield And Wakefield. *WIT Transactions on Ecology and the Environment*, 184, 163–173.

BIS. (2010). SME Statistics for the UK and Regions 2009. Enterprise Directorate Analytical Unit, Department for Business Innovation and Skills (BIS).

BIS. (2015). Business population estimates for the UK and regions 2015. Sheffield: Enterprise Directorate Analytical Unit, Department for Business, Innovation and Skills (BIS).

Blackburn, R. A. & Smallbone, D. (2008). Researching Small Firms and Entrepreneurship in the U.K.: Developments and Distinctiveness. *Entrepreneurship Theory and Practice*, 32(2), 267–288.

BMG Research. (2011). Cumbria Business Survey 2010 – Research report. Cumbria: Cumbria Intelligence Observatory.

Bowker, P., Escarameia, M. & Tagg, A. (2007). Improving the flood performance of new buildings: Flood resilient construction. London: Department for Communities and Local Government.

Climate Ready. (2014). *Business and Service theme* [Online]. Climate Ready, Environment Agency. Available: http://thecccw.org.uk/wp-content/uploads/2015/02/Climate-Ready-businesses-brief-present-and-future-work-areas-Apr-15.pdf [Accessed 10/09/2017]

Companies Act. (2006). *(c.46),* London, HMSO.

Crichton, D. (2006). *Climate Change and its effects on Small Businesses in the UK,* London, AXA Insurance UK.

Cumbria County Council. (2010). *Business and economy* [Online]. Cumbria: Cumbria County Council,. Available: http://www.cumbria.gov.uk/floods/oneyearon/businessandeconomy.asp [Accessed 15/10/2011].

Cumbria Intelligence Observatory. (2010). Cumbria Floods November 2009: An Impact Assessment Cumbria: Cumbria Intelligence Observatory.

Defra. (2008). Consultation on policy options for promoting property-level flood protection and resilience. London: Department for Environment, Food and Rural Affairs.

Defra. (2011). Understanding the risks, empowering communities, building resilience: The national flood and coastal erosion risk management strategy for England. London: The Stationery Office.

Douglas, I., Garvin, S., Lawson, N., Richards, J., Tippett, J. & White, I. (2010). Urban pluvial flooding: a qualitative case study of cause, effect and nonstructural mitigation. *Journal of Flood Risk Management,* 3(2), 112–125.

Environment Agency. (2009a). Cumbria 2009 Floods: Lessons Identified Report. London: Environment Agency.

Environment Agency. (2009b). Flooding in England: A national assessment of flood risk. Bristol: Environment Agency.

European Commission. (2006). The New SME Definition: User guide and model declaration. European Commission.

Evans, E., Ashley, R., Hall, J., Penning-Rowsell, E., Saul, A., Sayers, P., Thorne, C. & Watkinson, A. (2004). Foresight. Future flooding. Scientific summary: Volume I – Future risks and their drivers. London: Office of Science and Technology.

Federation Of Small Businesses. (2015). Severe weather: A more resilient small business community. London: Federation of Small Businesses.

Gissing, A., Molino, S. & Edwards, G. (2005). Business floodsafe – A toolkit for flood preparedness, response and recovery. Fourth Victorian Flood Management Conference, October 11–14, 2005 2005 Shepparton, Victoria. 1–9.

Gunningham, N. (2002). Regulating Small and Medium Sized Enterprises. *J Environmental Law*, 14(1), 3–32.

Hallberg, K. (2000). *A Market-oriented Strategy for Small and Medium Scale Enterprises*, Washington, World Bank.

HM Government. (2010). Securing Britain in an age of uncertainty: The strategic defence and security review. Norwich: The Stationery Office.

ICE. (2001). Learning to live with rivers – Final report of the ICE's presidential commission to review the technical aspects of flood risk management in England and Wales. London: Institution of Civil Engineers.

Ingirige, B., Jones, K. & Proverbs, D. (2008). Investigating SME resilience and their adaptive capacities to extreme weather events: A literature review and synthesis *Proceedings of the Conference on Building Education and Research (BEAR 2008)*. Kandalama, Sri Lanka.

Ingirige, B. & Russell, R. (2015). Investigating SME resilience to flooding – the Braunton report. Salford: University of Salford.

Ingirige, B. & Wedawatta, G. (2014). Putting policy initiatives into practice: Adopting an "honest broker" approach to adapting small businesses against flooding *Structural Survey*, 32(2), 123–139.

International Institute For Sustainable Development. (2004). Issue Briefing Note: Small and Medium-Sized Enterprises. *The ISO and Corporate Social Responsibility* [Online]. Available: http://inni.pacinst.org/inni/corporate_social_responsibility/standards_sme.pdf [Accessed 18/07/08].

IPCC. (2007). Climate Change 2007: The Physical Science Basis. Contribution of Working Group I to the Fourth Assessment Report of the Intergovernmental Panel on Climate Change. *In:* Solomon, S., D. Qin, M. Manning, Z. Chen, M. Marquis, K.B. Averyt, Tignor, M. & Miller, H. L. (eds.). Cambridge and New York Cambridge University Press.

Jha, A. K., Bloch, R. & Lamond, J. (2012). *Cities and Flooding: A Guide to Integrated Urban Flood Risk Management for the 21st Century*, Washington, The World Bank.

Kirchhoff, B. A. (1994). Entrepreneurship and dynamic capitalism: The economics of business firm formation and growth. Westport: Praeger Press.

Kreibich, H., Muller, M., Thieken, A. H. & Merz, B. (2007). Flood precaution of companies and their ability to cope with the flood in August 2002 in Saxony, Germany. *Water Resource Research*. 43(3), 1–15.

Kreibich, H., Seifert, I., Thieken, A. H. & Merz, B. (2008). Flood precaution and coping with floods of companies in Germany *In:* Proverbs, D., Brebbia, C. A. & Penning-Rowsell, E. (eds.) *Flood recovery, innovation and response*. Southampton: WIT Press.

Lamond, J., Bhattacharya-Mis, N., Chan, F., Kreibich, H., Montz, B., Proverbs, D. & Wilkinson, S. J. (2017). Flood risk mitigation and commercial property advice: an international comparison. London: Royal Institution of Chartered Surveyors.

Lauder, D., Boocock, G. & Presley, J. (1994). The System of Support for SMEs in the UK and Germany. *European Business Review*, 94(1), 9–16.
Lukacks, E. (2005). The Economic Role of SMEs in World Economy, Especially In Europe. *European Integration Studies*, 4(1), 3–12.
Marks, D. & Thomalla, F. (2017). Responses to the 2011 floods in Central Thailand: Perpetuating the vulnerability of small and medium enterprises? *Natural Hazards*, 87(2), 1147–1165.
McDaniels, T, Chang, S., Cole, D. & Longstaff, H. (2008). Fostering resilience to extreme events within infrastructure systems: Characterizing decision contexts for mitigation and adaptation, Global Environmental change, 18(2), 310–318
MCEER. (2005). White paper on the SDR Grand Challenges for disaster reduction, MCEER, Buffallo.
OECD. (2000). *OECD small and medium enterprise outlook: 2000 edition*, Paris, Organisation for Economic Co-operation and Development (OECD)
Paton, D. (2007). Measuring and monitoring resilience in Auckland. *GNS Science Report 2007/18*. Institute of Geological and Nuclear Sciences Limited.
Petts, J., Herd, A. & O'heocha, M. (1998). Environmental Responsiveness, Individuals and Organizational Learning: SME Experience. *Journal of Environmental Planning and Management*, 41(6), 711–730.
Pitt, M. (2008). The Pitt Review – Learning Lessons from the 2007 floods. London: Cabinet Office.
Pivot, J. & Martin, P. (2002). Farms adaptation to changes in flood risk: a management approach. *Journal of Hydrology*, 267(1–2), 12–25.
Robbins, D. K., Pantuosco, L. J., Parker, D. F., & Fuller, B. K. (2000). An empirical assessment of the contribution of small business employment to US state economic performance. *Small Business Economics*, 15(4), 293–302.
Runyan, R. C. (2006). Small Business in the Face of Crisis: Identifying Barriers to Recovery from a Natural Disaster. *Journal of Contingencies and Crisis Management*, 14(1), 12–26.
Stern, N. (2007). The economics of climate change: The Stern review. Cambridge: Cambridge University Press.
Tickner, L. (2011). Case Study: The Cockermouth flood recovery strategy. Presentation in Civil Contingencies 2011 conference, 18/01/2011 2011 London.
Tierney, K. (1995). Impacts of recent disasters on businesses: The 1993 midwest floods and the 1994 northridge earthquake. *Preliminary Paper 230* [Online]. Available: http://www.udel.edu/DRC/preliminary/230.pdf [Accessed 10/08/08].
Tilley, F. & Tonge, J. (2003). Introduction. *In:* Jones, O. & Tilley, F. (eds.) *Competitive Advantage in SMEs : Organising for Innovation and Change* UK: Hoboken, NJ John Wiley & Sons, Ltd.

UNDP. (2013). Small Businesses: Impact of Disasters and Building Resilience: Analysing the vulnerability of Micro, Small, and Medium Enterprises to natural hazards and their capacity to act as drivers of community recovery. *Background Paper prepared for the Global Assessment Report on Disaster Risk Reduction 2013*. Geneva: UNISDR.

UNISDR. (2004). Living with risk: A global review of disaster reduction initiatives. New York and Zurich: United Nations Office for Disaster Risk Reduction.

UNISDR. (2005). Know risk. Leicester and Geneva: Tudor Rose and United Nations Office for Disaster Risk Reduction.

Wedawatta, G. & Ingirige, B. (2012). Resilience and adaptation of SMEs to flood risk. *Disaster Prevention and Management,* 21(4), 474–488.

Wedawatta, G., Ingirige, B. & Proverbs, D. (2011). Adaptation to flood risk: the case of businesses in the UK. *International Conference on Building Resilience incorporating* 7th Annual International Conference of International Institute for Infrastructure, Renewal and Reconstruction (IIIRR). Kandalama, Sri Lanka.

Wedawatta, G., Ingirige, B. & Proverbs, D. (2012). Impacts of flooding on SMEs and their relevance to Chartered Surveyors. London: RICS Education Trust.

Wiseman, J. & Parry, E. (2011). Cumbria Business Survey 2010 – Research report prepared for Cumbria Intelligence Observatory. Birmingham: BMG Research.

Yoshida, K. & Deyle, R. E. (2005). Determinants of Small Business Hazard Mitigation. *Natural Hazards Review,* 6(1), 1–12.

CHAPTER 12

How Resilience Discourses Shape Cities: The Case of Resilient Rotterdam

Anique Hommels

Introduction: STS, the City and Resilience

Cities can be considered as enormous sociotechnical systems—as hybrid "assemblages" of social and material elements (Farias and Blok 2017). Science and Technology Studies (STS) scholars have argued that in cities, the social and the technical are co-constituted and that urban development can be seen as a process of sociotechnical change (Hommels 2005; Aibar and Bijker 1997). STS is an interdisciplinary field that studies how science and technology shape the world around us but also how society influences developments in science and technology (Felt et al. 2017). Studying cities from an STS perspective implies attention to the materiality of the city—its infrastructures and buildings—as well as to the social, political and economic organization of the city, human experiences, urban practices and imaginations.

Technology is all around us and cities can be seen as exemplars of living in a technological culture. STS scholars have argued that vulnerability is

A. Hommels (✉)
Faculty of Arts and Social Sciences, Maastricht University,
Maastricht, The Netherlands
e-mail: a.hommels@maastrichtuniversity.nl

S. Amir (ed.), *The Sociotechnical Constitution of Resilience*,
https://doi.org/10.1007/978-981-10-8509-3_12

inherent to our modern high-tech societies (Bijker et al. 2014). This argument seems to apply to cities in particular. Over the past decade, urban scholars have argued that cities have become highly vulnerable. Mark Pelling, for instance, (2003) has argued that: "cities are increasingly becoming the locus of risk" (p. 14). And the neo-Marxist geographer David Harvey stressed that "cities are vulnerable forms of human organization. (....) In recent times, the extraordinary growth of cities through the world, seems set to override catastrophes, losses, indignities and woundings, no matter whether externally visited or self-inflicted" (Harvey 2003, p. 25).

However, cities have also been argued to be among the world's most durable and resilient constructions (Haas et al. 1977). Cities are likely to endure despite natural disasters or other kinds of assaults. They are usually rebuilt after a catastrophe and a complete abandonment or relocation of a city in a post-disaster period is very rare. Urban geographer Nigel Thrift argues that "it has become increasingly clear that cities are actually extraordinarily resilient: they routinely bounce back from accident and disaster" (Thrift 2005, p. 343). In a similar vein, urban planning scholars Lawrence Vale and Thomas Campanella (2005) argue that "although cities have been destroyed throughout history – sacked, shaken, burned, bombed, flooded, starved, irradiated and poisoned – they have, in almost every case, risen again like the mythic phoenix" (p. 3).

These examples show the ambiguous character of the city as both a vulnerable and resilient or durable system. Depending on one's perspective, a city can be seen as a risk-prone place but also as a strong, innovative breeding ground. Vulnerable places may provide opportunities for innovation because of the flexibility and openness they seem to foster (Bijker 2006). Thus, STS researchers have argued that vulnerability is not only or purely negative. Some vulnerability is even necessary to allow for learning and creative adaptation in a society: "Once properly addressed, such vulnerability with accompanying coping mechanisms may yield a more flexible and resilient society than one that tries to avoid all vulnerabilities" (Hommels et al. 2014).

This ambiguity of cities and the current challenges related to urban sustainability have resulted in a widespread attention for the notion of resilience in both urban scholarship and in urban planning and policy making (Vale 2014). Resilience is a "slippery" term. The concept has now been used in a variety of disciplines, ranging from ecology and engineering to planning and social sciences. It has such a flexible meaning that it lends itself easily to reinterpretation and integration in a diversity of academic fields. Therefore, the concept has the potential of a being "bridging concept" that can support interdisciplinary cooperation between diverse scholarly disciplines in the social sciences, humanities and engineering.

Over the past decade, resilience has increasingly become adopted by planners and urban scholars as a way to describe "the ability of cities to respond to threats to its system" (Vale 2014). In urban scholarship, the term can refer to a city's responsiveness to sudden changes like disasters, but also to gradual and more predictable developments (such as de-industrialization, economic decline or urban shrinkage). Yet, similar to other fields, the current status of resilience as a concept in urban research is volatile and contested. On the one hand, scholars argue that the notion of resilience offers a paradigm shift to planning theory and practice (Shaw 2012). On the other hand, some scholars are quite critical about the application of the concept: "There is a sneaking suspicion that much of what has been recently labelled 'resilience' is 'old wine in new bottles'" (Weichselgartner and Kelman 2015, p. 259). They point out that it is important to learn from scientific research that has been done and that it makes no sense to try describing extensively discussed phenomena with newly invented terms. Another problem is that, since the notion of resilience has such diverse interpretations and applications, it is difficult to put it into practice in an urban policy making context (Spaans and Waterhout 2016).

This chapter analyzes a concrete case of a European "resilient city" in development to see how resilience discourses shape cities. The Dutch city of Rotterdam, a member of the world-wide 100 Resilient Cities initiative of the Rockefeller Foundation, will be studied as an example of the attempted embedding of particular meanings of resilience. Rotterdam joined the 100 Resilient Cities initiative in 2013 and launched its Resilient Rotterdam strategy in the spring of 2016. This chapter analyzes this strategy in detail, studies the notions of resilience underlying it, and uses STS literature to critically reflect on some of the tensions and challenges involved in urban resilience discourses and their application.

To better grasp the way resilience has been conceptualized in the context of cities and urban planning, I will first discuss the scholarly debate on urban resilience in more detail. Then, I will analyze the specific way in which the 100 Resilient Cities initiative has interpreted urban resilience and how the city of Rotterdam appropriated this discourse in its resilience strategy. I will conclude this chapter by discussing some of the tensions and challenges of Rotterdam's resilience thinking, taking inspiration from STS literature on cities, disasters and vulnerability.

Urban Resilience as a Theoretical Construct

Two important approaches in resilience research, engineering resilience and ecological resilience, can be considered as the roots of the current discussions on urban resilience. The ecologist Holling (1973) introduced the notion of engineering resilience. In this approach, resilience is defined as the capacity of a system to bounce back after a severe disruption. The system's ability to find a new equilibrium and the time that is needed to do so is seen as the best marker of the system's resilience. In ecological resilience, the focus is on the ability of eco-systems to persist despite pressures on their functioning. This perspective emphasizes the management of change (Healy and Mesman 2014).

Urban planning scholars have criticized such equilibrium or "bouncing back" approaches. Cities might get reconstructed without being fully recovered (Campanella 2006), and urban resilience entails far more than urban rebuilding (Campanella 2006; Vale and Campanella 2005). Moreover, a return to "normal state" or prior equilibrium is not even always desirable, as Davoudi (2012) argues. The example of Hurricane Katrina in 2005 shows that "it not only destroyed the physical fabric of New Orleans, but also revealed social processes which many people did not find as the acceptable, pre-disaster normal to which they wanted to return. On the contrary, what was aspired to was a 'new normal' in social, economic and political terms" (Davoudi 2012, p. 302). Urban scholars stress the normativity involved in the question to which "state" a system best returns to. Another critique is that much of the literature in this field focuses on huge, large-scale, big impact disasters. There is less attention to "gradual, small and cumulative changes" (Davoudi 2012, p. 302).

In response to equilibrium models of resilience, an alternative discourse has developed under the label of "non-equilibrium approaches," or "bouncing forward discourses" (Shaw 2012). Some scholars consider the non-equilibrium model to be better suited to the urban context (Vale 2014, p. 192). As this discourse frames resilience as the ability of a system to adapt to changing external or internal circumstances, it is claimed to fit better with urban planning because "it is more dynamic and evolutionary" (Pickett et al. 2004, p. 373). The non-equilibrium paradigm focuses attention on processes and dynamics rather than on states and structures (Pickett et al. 2004, p. 381). It links to a research agenda within planning theory that has more eye for the implications of the non-linear dynamics of ecosystems (Wilkinson 2012, p. 149).

The "evolutionary resilience" (Davoudi 2012) and the "socio-ecological systems" (SES) approaches (Evans 2011; Wagenaar and Wilkinson 2013; Wilkinson 2012) were introduced to counter the shortcomings of equilibrium oriented approaches of resilience. In contrast to views on resilience that see systems as mechanical, objective, orderly and more or less predictable, the evolutionary view of resilience acknowledges their complex, chaotic, uncertain and unpredictable features. Crucially, resilience, in this perspective, is seen as a process, not as a state of being that can be assessed at one moment. Moreover, attention is paid to the role of institutions, leadership, social capital and social learning. Similarly, the SES approach has been developed to emphasize resilience and adaptive learning as strategies to enhance urban sustainability in, for instance, the case of climate change (Evans 2011, p. 224; Wardekker et al. 2010).

Following the "bouncing forward" discourse, the notion of urban resilience has become increasingly applied to urban policy. According to Lawrence Vale (2014, p. 191, my emphasis) "resilience is, simultaneously, a *theory* about how systems can behave across scales, a *practice* or proactive approach to planning systems that applies across social spaces, and an *analytical tool* that enables researchers to examine how and why some systems are able to respond to disruption." Nowadays, resilience discourses have been adopted by a wide range of international urban projects as well as national and urban policy agendas (Evans 2011). In urban policy, resilience can be *pro-active* (anticipatory) and *reactive* (urban planners get involved after a disaster has occurred) (Vale 2014). Moreover, the normative and political dimensions of urban resilience have received more attention lately; human intervention should be more prominent on the research agenda of urban scholars, as should questions of justice and fairness (as in the example of Hurricane Katrina) (Davoudi 2012). Thus, current views on urban resilience tend to emphasize the non-linear, complex and unpredictable character of urban systems and increasingly pay attention to the normative and political aspects involved in dealing with urban resilience.

In a recent article, STS scholars Amir and Kant (2018) introduced the notion of "sociotechnical resilience." They criticize the engineering perspectives on resilience as well as approaches that focus on resilience from a human factors, risk or disaster perspective. They argue that these perspectives fall short in fully addressing the challenges of sociotechnical systems, since the sociotechnical hybridity of these systems is not completely acknowledged and understood in these approaches. Instead, they propose a hybrid interpretation of resilience, understanding both people and technologies as social constructs and systems as both social and technical. The notion of "transformability" is at the core of their conceptualization of sociotechnical resilience. This refers to "the ability of a sociotechnical system to shift from one configuration to another in response to shock and disruption" (Amir and Kant 2018, p. 8). They further conceptualize sociotechnical resilience by focusing on three aspects: (1) informational relations (how information is organized and managed to make sure that sociotechnical systems can function without disruptions), (2) sociomaterial structures (referring to the entanglement of social and material structures) and (3) anticipatory practices (activities aimed at anticipating future adverse events and alternative scenarios). Understanding the city as a hybrid sociotechnical system aligns well with this conceptualization of sociotechnical resilience.

Lu and Stead (2013) argue that "the notion of resilience is important in cities for two reasons: it provides a new way of framing and responding to uncertainty and vulnerability in spatial planning and urban development; and it offers an alternative paradigm for developing strategies and approaches to deal with large-scale social, environmental and economic changes in cities" (p. 211). Yet, urban resilience is not only influential in academic discourses but is becoming increasingly visible and influential in policy documents across the globe. One of the major policy initiatives in this respect is the 100 Resilient Cities initiative of the Rockefeller Foundation. In the next section, I will analyze how urban resilience is framed in the 100 Resilient Cities initiative.

The 100 Resilient Cities Initiative and Its Perspective on Resilience

The Rockefeller Foundation was established in 1913 and has been active in urban policy programs since the 1950s when it launched its Urban Design Studios Program. In its celebration of its 100-year anniversary, the Rockefeller Foundation launched the 100 Resilient Cities initiative, "a non-profit, dedicated to helping cities around the world become more resilient to the physical, social and economic challenges of the 21st century" (The Rockefeller Foundation 2017, p. 12). The mission of the Rockefeller Foundation is to "catalyze an urban resilience movement" (p. 12).[1]

The Foundation does this, for instance, by establishing partnerships with cities around the world (they are now present on six continents in 48 countries) and by funding a two-year appointment of a so-called Chief resilience officer (CRO) in the member cities. Member cities have to develop a resilience strategy—a process led by the CRO and supported (in terms of training, capacity building, technical assistance and access to resources) by the Rockefeller Foundation. So far, 80 CROs have been appointed. The Rockefeller Foundation also built what they call a "Platform of partners," companies, industries and innovators from private and non-profit sectors who can deliver technical solutions for the problems cities are coping with. "Cities often lack access to the tools or technical assistance they need to design and implement solutions to their resilience challenges; sometimes they don't even know what kind of support they need" (p. 12). The Rockefeller Foundation tries to influence global actors (thought leaders, policy makers, financial institutions) to get financial support for the implementation of resilience projects in member cities. So far, they were able to raise 535 million dollars for this.

The report "Cities taking Action" was published in July 2017 and tries to take stock of the efforts to enhance urban resilience within the 100 Resilient Cities initiative of the Rockefeller Foundation. Started in 2013, the report argues, "the importance of cities to the 21st century was gaining recognition, but urban resilience and the urgent need to develop it was not at the forefront of conversations about our global future. Today it is" (Letter from the President, Michael Berkowitz). Furthermore, the Letter from the President states that this report marks the transition from planning for resilience to actually building urban resilience. The report emphasizes that cities are of crucial importance since by 2016 more than 55% of the world's population was living in cities, a number that is predicted to rise to 70% by 2050. "But the impacts of climate change, aging infrastructure, population growth and mass migration, and social and economic inequity, are all disproportionally borne by cities today" (p. 8). The report argues that whereas national governments are often slow in making decisions, local leaders do not have the luxury for that: "they must meet the day-to-day needs of their residents, and ensure any investment returns tangible benefits...The strategies cities create for their futures, and the decisions they make on what to prioritize, will reverberate globally, with the potential to affect the lives of billions of people" (p. 8).

The report characterizes the twenty-first century by highlighting three important trends: urbanization, globalization and climate change. It is argued that "these conditions require new models of governance" (p. 10). Building resilience is necessary, it is argued, to allow cities to prosper under these circumstances. Urban resilience is defined as "the capacity of individuals, communities, institutions, businesses, and systems within a city to survive, adapt and grow no matter what kinds of chronic stresses and acute shocks they experience" (p. 10). Acute shocks are "sudden, sharp events that threaten a city, such as earthquakes, disease outbreaks, or terrorist attacks" (p. 10). Chronic stresses are, for example, high unemployment or inefficient transportation systems, institutional racism, or poor macroeconomic conditions. They "weaken the fabric of a city over time and exacerbate shocks when they inevitably occur" (p. 10). Often shocks and stresses coincide. Therefore, the report says, it is necessary to plan "holistically" "so that a city is prepared for whatever shocks and stresses may arise" (p. 10). The example of Paris shows how quickly urban vulnerabilities may change: when they applied for the 100 Resilient Cities network in 2014, the city focused on flooding and heat waves. However, since the terrorist attacks of 2015 the city had to refocus its priorities and develop more "holistic strategies for strengthening the city, ones that would build inclusive, cohesive, and prepared communities" (p. 11).

The report criticizes the current way of working in many cities which is characterized by "a siloed approach" in which, for instance, work on disaster recovery, sustainability, livelihoods, and infrastructure development are split up in different departments. "This may be an efficient way to structure the work of a city, but it is not the most effective way. Cities are systems, not silos" (p. 11). If cities want to take planning for resilience seriously, it is necessary to opt for an integrated and inclusive approach, according to the report.

The breaking down of siloes is one of the tasks of the CRO and one of the key tools to do so is the development of a resilience strategy. The common task of developing the city's resilience strategy is already part of building resilience by "entrenching a practice of integrated, inclusive, risk-aware and forward looking planning" (p. 18). The idea is that citizens and vulnerable groups in the city in particular are invited to join the process of developing the resilience strategy: "Deep and meaningful dialogue with citizens and stakeholders is a critical aspect of creating a Resilience Strategy.... Because the poor and vulnerable are also often disproportionately impacted by shocks and stresses, their perspective must inform the creation of the policies and programs that will affect them" (p. 19). The final goal is to "collectively catalyze a global movement to build urban resilience" (p. 19).

Four years after its initial start in 2013, the next step in the process is to move from making plans and building strategies towards implementation: "The next era of our partnership with these dynamic cities and global leaders will be focused on implementation. It is one thing to talk about transformation, it's another to design, resource and implement actions and initiatives that actually deliver systemic change" (p. 64). In the next section, I will look into one of the cities in this network: the Dutch city of Rotterdam. Rotterdam has been a participant in this network since 2013. Analyzing its Resilience Strategy[2] will make clear how this city tries to embed resilience thinking in the urban fabric and some of the challenges and tensions involved in this effort.

Resilient Rotterdam: How Resilience Thinking Takes Shape

The Dutch city of Rotterdam has around 600,000 residents, and is the largest freight port in Europe. Around 80% of the city lies below sea-level. The city and its surroundings are protected by a system of dikes, closure dams and storm surge barriers (Molenaar et al. 2013). The city claims to have "a long tradition of developing resilience" (Burgemeester en Wethouders van Rotterdam 2016, p. 1). The Second World War bombings in May 1940 that devastated a large part of the historic center of Rotterdam are often mentioned as exemplary for the way the city copes with disaster and shock. "The Rotterdam society turned out to be robust in dealing with drawbacks: while the occupation still lasted, cleaning up the debris started and plans were made for urban renewal. This attitude was and is characterized by flexibility and creativity: a new innovative city center emerged and the initial chilliness and emptiness are turned into an inclusive City Lounge in 2015" (Burgemeester en Wethouders van Rotterdam 2016, p. 1).

In a public speech in June 2017, mayor of Rotterdam Ahmed Aboutaleb defined resilience as "the building of dikes, not only to protect us from water, but from everything." He argues that Rotterdam has "small splashes of fuel everywhere and we only have to wait until they are set on fire." Resilience is the only policy domain that brings him sleepless nights, he says.[3] Thus, the way resilience is framed in Rotterdam sounds quite paradoxical: on the one hand, the city is already very resilient if we look at past events that the city successfully recovered from (e.g., the Second

World War bombings). On the other hand, the city is not resilient yet, as it has to deal with multiple challenges (the metaphorical splashes of fuel) that threaten its future.

Because of its vulnerable geographical location, Rotterdam actively developed climate adaptation strategies over the past decades. These efforts were framed in terms of enhancing urban sustainability. When the members of the City Resilience team of Rotterdam were approached by the Rockefeller Foundation in 2014, they thought that it would be possible to re-frame the climate adaptation challenges into a resilience framework.[4] With this goal in mind, they decided to join the 100 Resilient Cities network (interview member CRO team Rotterdam). But soon it became clear, that the aims of the Rockefeller Foundation were much broader than climate adaptation and that Rotterdam had to include other areas to make its strategy fit with the aims of enhancing resilience. The specific challenges that are highlighted in the Rotterdam Resilience Strategy are the following: (1) social cohesion and education, focusing on the individual and societal level; (2) clean energy transition; (3) climate adaptation by implementing clever water management approaches and better understanding cascading failures; (4) cyber use and security, by making the port of Rotterdam and the city more cyber secure (through collaboration, knowledge sharing and awareness); (5) critical infrastructure, especially Rotterdam's underground infrastructure needs to become more flexible and responsive to emergencies; (6) changing urban governance by connecting networks of locals, individuals and businesses; and (7) anchoring resilience in the city (Gemeente Rotterdam 2016). The time horizon within which the resilience vision has been developed is 2030. By then, the city should have become more socially balanced, cleaner, more climate resilient and cyber secure, more socially cohesive and networked and with a more flexible and disaster responsive underground infrastructure.

In the remainder of this chapter, I will focus on three core principles[5] of the Rotterdam Resilience Strategy: (1) that resilience thinking requires an integrated approach, (2) that it requires innovation and flexibility and (3) that it is an inclusive approach involving a wide range of stakeholders. I will argue that the way in which these three principles are currently performed in Rotterdam does not cohere with recent STS insights on urban hybridity and sociotechnical resilience.

The first principle of "integration" clearly refers to the holistic perspective and the notion of "breaking down siloes" in the Rockefeller report. In Rotterdam, implementing this perspective is the task of Arnoud

Molenaar, Rotterdam's CRO. According to him, the key goal of Rotterdam's Resilience Strategy is to "make resilience common practice in our city and part of our DNA" (Gemeente Rotterdam 2016, p. 11). And: "The goal of our resilience strategy is to integrate the seven aspects of resilience into our thinking and actions in a more focused and systematic manner" (p. 42). An integrated approach in this context means that one should "look beyond just one field or economic sector" (p. 122). Moreover, it means "actually applying the resilience lens in day-to-day thinking and actions" and "across all levels: city, district, street or building" (p. 122). The integrated approach should in the end result in a deep anchoring of the concept of resilience in the city of Rotterdam. It aims to create a movement "that will bring resilience into the mainstream, and thus integral to all our thoughts and actions" (p. 125).

The second core value of the resilience strategy of being innovative and flexible is also very prominent in Rotterdam's resilience thinking: "Building a resilient city will demand an open mind and new ways of thinking…Building a resilient city means built-in flexibility, freeing up funds, thinking beyond our own interests, and looking at the real questions that we face as a city, rather than just settling for solutions that are already available" (p. 40). Achieving resilience asks for some specific innovations in the city such as enhancing the environmental friendliness of the port and its resilience to cyber-attacks. Moreover, by 2025, it is argued, Rotterdam should be "a resilient city that is able to withstand the effects of climate change, such as extremely heavy rainfall, rising temperatures and rising water levels" (p. 86). To make such innovations possible, it is important to be flexible and responsive in governance, the strategy argues: "It will therefore be necessary to respond to the changing characteristics of our system of governance and to enable other actors to play a role in this" (p. 114).

Third, an "inclusive approach" means that a diversity of stakeholders need to be involved in making cities more resilient. According to the Rockefeller Foundation, "being inclusive" refers to "the need for broad consultation and engagement of communities, including the most vulnerable groups" (Rockefeller Foundation and Arup 2013). It also relates to the idea of a shared vision of resilience among the various stakeholders.

As this section shows, the urban resilience discourse of Rotterdam is infused with terms such as integration, flexibility, inclusion and innovation. The wide-ranging ambitions and the multiplicity of initiatives included in the resilience strategy give the impression of a very rich and diverse set of approaches to make the city more resilient. However, in practice, resilience discourses may be difficult to translate into concrete policies: "To accept resilience thinking boils down to embracing change and embracing complexity. It is a big picture approach, which can theoretically encompass everything that exists" (Martin-Breen and Anderies 2011, p. 52). Therefore, some scholars have pointed out that this begs the question whether there are policy domains or approaches that are *not* relevant to making the city resilient. As Spaans and Waterhout (2016) argued, "resilience planning and preparedness may require a selective rather than a holistic approach... [and] each type of resilience may require a different approach and strategy" (p. 7). It is also a political question of which elements or functions of a city are selected to become resilient. And cities can be quite obdurate, both institutionally and physically. How does the resilience approach of flexibility, integration, inclusion and innovation fit the reality of the existing, obdurate urban sociotechnical ensemble? In the next section, I will confront the resilience principles of Rotterdam with some thoughts and ideas that were developed in STS (Science, Technology and Society studies) over the past few years, and argue that more work needs to be done to make the Rotterdam resilience strategy live up to its own key principles.

STS Applied to Rotterdam's Resilience Strategy

STS scholars have emphasized the hybrid character of cities as sociotechnical systems or assemblages (Farias and Blok 2017; Hommels 2005) and proposed to conceptualize resilience in terms of "sociotechnical" resilience (Amir and Kant 2018). Lu and Stead (2013) have argued that possibly, the notion of resilience may imply a shift from hard engineering approaches (where infrastructure is used to deal with risks and threats) to softer approaches that focus on coping with the effects of disasters (p. 211). Yet, a shift from engineering approaches towards softer perspectives on resilience does not automatically result in a more heterogeneous vision of resilience. As the case of Rotterdam has shown, resilience is not approached as "an emergent property produced through a process whereby the structure of social organizations and the arrangement of technical systems are constantly intertwined" (see Introduction by Amir). Rather, both the structure of the project in more technical and more social sub-projects, and the institutional set-up of the municipality, make it difficult to break down the "siloes" of technology and society.

Although a major effort has been made to break down these siloes and develop an integrated approach towards resilience, arguably, the Rotterdam Resilience Strategy has not yet reached its aim of integrated thinking. What is missing is a reflection on the hybrid character of cities as both social and material. As the editor of this book argued: "Rather than seeing resilience as being fully reinforced by social connections or an outcome of inherent strength of infrastructures, we see resilience as an embedded feature that comes out of a hybrid realm where individuals and communities are blended with the materiality of technology" (Introduction, p....). In a similar vein, Healy and Mesman argue "while conceptualizations of resilience tend to be level-bound [micro/meso/macro], concentrating on either the micro-, meso-, or macro-level, achieving resilience in practice usually requires coordination involving not only people, but also things across all these levels" (p. 174).

Yet, despite the broad and integrative approach that characterizes resilience thinking in Rotterdam, it is not so hybrid that it does justice to all the complexities of the urban sociotechnical reality. The materiality of the city comes into play when underground infrastructure challenges are discussed and when measures for mitigating climate change are addressed. The urban social fabric and the role of communities gets attention when resilient neighborhoods and the "we" society are discussed. Yet, an awareness of the intricate interrelationships between society and technology is not apparent from the strategy. Furthermore, the way the city is governed, by specific departments and portfolios, makes it institutionally difficult to establish a more integrated approach.[6] As one of the CRO team members emphasizes, it is very difficult to translate resilience thinking into concrete projects or approaches. For him, it is equally important to change the institutional organization of the municipality and embed resilience thinking in all municipal layers. This can make it easier to implement concrete resilience projects in the future.

The specific discursive framing of urban resilience in Rotterdam has already materialized in some urban planning decisions and land-use management and will continue to do so in the future (Lu and Stead 2013, p. 211). However, the embedding of these sociomaterial changes may be complicated due to the obduracy of the existing urban structures (Hommels 2005). Furthermore, there seems to be a tension between the idea of radical urban innovation and change, and the more conservative idea of adaptive capacity (and bouncing back) in resilience thinking.

This tension is addressed in the work of STS scholars Healy and Mesman (2014) who discuss the difference between literature on the adaptive management of socio-ecological systems and resilience in transition management. They argue that "adaptive resilience" can be an impediment to the aims of transition management (facilitating radical, systemic, sociotechnical transformation p. 171). Yet, there are also attempts at reconciling adaptive management and transition management as Healy and Mesman claim. Both use participatory methods to identify the concerns of local stakeholders. They also describe system boundaries and the institutional context: "Reconciling adaptive management with transition management, then, necessitates an inclusive consultative process designed to normatively reconcile the ecological imperatives of the former with the sociotechnical imperatives of the latter, integratively engaging considerations, both human and non-human, at all levels" (p. 173).

A report in which the transition management perspective is applied to the resilient Rotterdam case shows that indeed the emphasis is on accelerating urban change rather than on "bouncing back" (Lodder et al. 2016). The report aims to support and accelerate the transition in the direction of a more resilient Rotterdam. They want to push the developments beyond small-scale experiments and scale up to a transition pathway: "To accelerate the transition, disruptive innovations need to be up-scaled, and those aspects of the current system that hamper resilience need to be scaled-down" (p. 43). In this case, it is important to distinguish between reactive and pro-active (anticipatory) approaches towards resilience (Vale 2014). The Rotterdam Resilience Strategy is a clear example of a pro-active approach whereas reactive approaches are more inclined to a bouncing-back vocabulary.

Both STS and urban scholars have emphasized the need to critically assess the normative and political implications of resilience thinking (Healy and Mesman 2014; Vale and Campanella 2005): "(…) tensions emerge because of inattention to normative considerations, most notably the resilience of what, for whom, and at what costs and/or tradeoffs? These are critical matters that any analysis or practically focused initiative concerned with resilience and vulnerability must address explicitly to be effective" (Healy and Mesman 2014, p. 173). A similar argument is advanced by "STS disaster" scholar Kim Fortun et al.: "Especially important for our argument here is the way resilience (…) works toward recovery and stability in ways that can reproduce previously entrenched disadvantage" (Fortun et al. 2017, p. 1017). They refer to the UN-approved "Hyogo

Framework" for disaster reduction, which sounds quite responsive in its aims, adopting a "bouncing back" vocabulary. "With such an orientation, the ways functional systems inevitably have and produce margins (and thus disadvantage) is easily obscured" (p. 1017).

Andy Stirling adds that aims such as advancing human well-being, social equity and environmental integrity are broad normative aims that require the inclusion of vulnerable groups themselves. This normativity has a temporal aspect: it is not fixed what the policy outcomes should be. "Presuming that the normativities of resilience are self-evident tends to assert hegemonic values at the expense of the less visible interests of more marginal groups" (Stirling 2014, p. 323). The Rotterdam Resilience Strategy tries to be inclusive. Yet, its resilience team also admits that involving vulnerable groups, such as homeless people, unemployed or very poor people, is not that easy. The resilience strategy is mainly shaped by other stakeholders such as housing corporations, research institutes and utility companies. Although the city tries to stimulate local initiatives of citizens, it is complicated to involve representatives of vulnerable groups in the development of the resilience strategy.

I have argued that STS and urban studies research highlighted some tensions in Rotterdam's resilience discourse that mainly pertained to the problem of bridging the boundary between social and technical resilience, the difficulty to embed divergent meanings of resilience in concrete trajectories for sociotechnical change, and the challenge of performing resilience in a policy that supports diversity and social inclusion. Table 12.1 summarizes the main tensions in Rotterdam's resilience thinking.

Conclusions: How Resilience Thinking Shapes the City

This chapter has shown that urban resilience discourses often stress the innovative, flexible, integrative and inclusive character of the city. These ambitions can be useful signposts for the future of cities. Yet, looking at the specific example of resilient Rotterdam from an STS perspective, some tensions and challenges in the discourse become visible.

First of all, the example of Rotterdam shows that both conceptually and institutionally, it is difficult to break down the existing siloes in cities. Looking at this with an STS lens shows that in the case of Rotterdam, social and technical resilience are perceived as quite distinct. This becomes

Table 12.1 Tensions in Rotterdam's resilience discourse

Core principles of the Rotterdam Resilience Strategy	*Related tensions*	*Implications for resilience and the city*
Integral	Institutional siloes difficult to break down Project structure distinguishes "technology-" and "society-"related resilience	Resilience challenge is not approached in its full complexity
Innovative and flexible	Difficult to implement, still hinging on old climate adaptation policies, not concrete enough Different interpretations of resilience are still being negotiated	Resilience thinking may be difficult to concretize/embed
Inclusive	Stakeholders are narrowly defined and most vulnerable groups are difficult to reach	Resilience may be become a definition of a limited group of people and therefore difficult to scale up

clear in the list of resilience-oriented initiatives where social resilience projects (such as the "we" society) are distinguished from infrastructure or climate related resilience projects. Moreover, institutionally (in the way the municipality is organized), it turns out to be difficult to break down boundaries between policy domains. However, resilience thinking potentially acts as a catalyzer towards a more integrated approach. And indeed, a more radical integrated approach in terms of sociotechnical resilience is necessary to approach the resilience challenge in all its complexity.

Second, the vocabulary of inclusiveness and integration may obscure the fact that normative choices are often left implicit in resilience strategies. Normative implications and trade-offs need to be made explicit (for example goals, aims and approaches need to be more clearly distinguished). Vulnerable groups need to be more included in framing resilience to avoid that the understanding of one dominant group (e.g., policy makers or local industry) dominates the process of policy development and implementation. One of the tensions in resilient city projects such as Rotterdam is that the aim is to be inclusive, but vulnerable groups cannot be reached.

Third, more attention is needed, now that the implementation phase is coming, for the translation to concrete results. Resilience is a broad concept and has such flexible meanings that it can be elusive. In Rotterdam, resilience is about enhancing social cohesion as well as climate adaptation, cyber security, protecting critical infrastructure and so on. Despite the attempts to broaden the scope of resilience in the city, it appears that when it comes to concretizing the notion in urban projects, the city's former emphasis on climate adaptation and sustainability is still dominant in the resilience discourse.

However, the current focus in Rotterdam seems to be more on institutionalizing a specific way of thinking in urban governance and in gathering critical mass for a resilience movement than on implementing concrete projects, or sociotechnical innovation. Rotterdam's Resilience Strategy is part of a longer-term municipal learning trajectory. It is not self-evident that the next city board (municipal elections will take place in 2018) will continue developing the resilience strategy. The current CRO team definitely intends to convince the city board that resilience thinking is beneficial for the city and the city's governance. But for that, they also need concrete results.

This chapter has shown how resilience thinking currently shapes cities all over the world, and what the political and normative implications of this movement can be. Analyzing the assumptions, principles and possible tensions underlying urban resilience strategies is one step. The next step is to study how these ideas, expectations and ambitions get embedded in the sociomaterial structures that cities are. Since resilience thinking has become so widespread, STS scholars have many ways to contribute to the analysis of these efforts.

Notes

1. See www.100resilientcities.org for more information on this initiative.
2. I analyzed the Rotterdam Resilience Strategy by studying key documents that were produced over the past three to four years: consultation documents, minutes of meetings of the steering group, reports (sometimes written by external partners of Rotterdam, such as the university or research institutes), leaflets, working plans, letters and so on. I also held a background interview with one of the members of the core team Resilient Rotterdam and I visited the "One year Resilient Rotterdam" event on June 26, 2017.

3. Public speech Mayor Ahmed Aboutaleb, June 26, 2017. One year Resilient Rotterdam event. Rotterdam.
4. Rotterdam's shift from sustainability to resilience thinking can also be understood from a political perspective. In 2014, the social democrats lost many seats in the city council of Rotterdam whereas this party has been most supportive of the sustainability agenda. A shift to resilience would make this agenda more appealing to the conservatives (Terenzi 2014).
5. The Rockefeller Foundation defined seven resilience principles. Resilient cities should be: reflective, resourceful, robust, redundant, flexible, inclusive and integrated. My analysis focuses on the last three principles.
6. Interview CRO team member Corjan Gebraad.

References

Aibar, E., & Bijker, W. E. (1997). Constructing a city: The Cerdà plan for the extension of Barcelona. *Science, Technology, & Human Values 22*(1), 3–30.

Amir, S. & Kant, V. (2018). Sociotechnical Resilience: A Preliminary Concept. Risk Analysis, 38: 8–16. https://doi.org/10.1111/risa.12816.

Bijker, W. E. (2006). The Vulnerability of Technological Culture. In H. Nowotny (Ed.), *Cultures of Technology and the Quest for Innovation*. New York: Berghahn Books.

Bijker, W. E., Hommels, A., & Mesman, J. (2014). Studying Vulnerability in Technological Cultures. In A. Hommels, J. Mesman, & W. E. Bijker (Eds.), *Vulnerability in Technological Cultures. New Directions in Research and Governance* (pp. 1–26). Cambridge, MA: MIT Press.

Burgemeester en Wethouders van Rotterdam. (2016). *Letter draft Resilience Strategy (concept Resilience Strategie)*. (19 May 2016). Rotterdam.

Campanella, T. (2006). Urban Resilience and the Recovery of New Orleans. *Journal of the American Medical Informatics Association, 72*(2), 141–146.

Davoudi, S. (2012). Resilience: A Bridging Concept or a Dead End? *Planning Theory & Practice, 13*(2), 299–333.

Evans, J. P. (2011). Resilience, ecology and adaptation in the experimental city. *Transactions of the Institute of British Geographers, `36*, 223–237.

Farias, I., & Blok, A. (2017). STS in the City. In U. Felt, R. Fouche, C. A. Miller, & L. Smith-Doerr (Eds.), *The Handbook of Science and Technology Studies* (pp. 555–581). Cambridge, MA: MIT Press.

Felt, U., Fouche, R., Miller, C. A., & Smith-Doerr, L. (2017). Introduction to the Fourth Edition of The Handbook of Science and Technology Studies. In U. Felt, R. Fouche, C. A. Miller, & L. Smith-Doerr (Eds.), *The Handbook of Science and Technology Studies (4th Edition)* (pp. 1–26). Cambridge, MA: MIT Press.

Fortun, K., Knowles, S. G., Choi, V., Jobin, P., Matsumoto, M., De la Torre III, P., ... Murillo, L. F. R. (2017). Researching Disaster from an STS perspective. In U. Felt, R. Fouche, C. A. Miller, & L. Smith-Doerr (Eds.), *The Handbook of Science and Technology Studies* (pp. 1003–1028). Camrbidge, MA: MIT Press.

Gemeente Rotterdam. (2016). *Rotterdam Resilience Strategy. Ready for the 21st century (Consultation document. Extended version)*. Retrieved from https://www.resilientrotterdam.nl/.

Haas, J. E., Kates, R. W., & Bowden, M. J. (Eds.). (1977). *Reconstruction following Disaster*. Cambridge, MA: MIT Press.

Harvey, D. (2003). The City as a Body Politic. In J. Schneider & I. Susser (Eds.), *Wounded Cities. Destruction and Reconstruction in a Globalized World* (pp. 25–44). Oxford: Berg.

Healy, S., & Mesman, J. (2014). Resilience: Contingency, Complexity and Practice. In A. Hommels, J. Mesman, & E. Bijker Wiebe (Eds.), *Vulnerability in Technological Cultures. New Directions in Research and Governance* (pp. 155–177). Cambridge, MA: MIT Press.

Holling, C. S. (1973). Resilience and stability of ecological systems. *Annual Review of Ecology and Systematics, 4*, 1–23.

Hommels, A. (2005). *Unbuilding Cities. Obduracy in Urban Sociotechnical Change*. Cambridge, MA: MIT Press.

Hommels, A., Mesman, J., & Bijker, W. E. (Eds.). (2014). *Vulnerability in Technological Cultures. New directions in research and governance*. Cambridge, MA: MIT Press.

Lodder, M., Buchel, S., Frantzeskaki, N., & Loorbach, D. (2016). *Richting een Resilient Rotterdam. Reflecties vanuit een transitie-perspectief.*

Lu, P., & Stead, D. (2013). Understanding the notion of resilience in spatial planning: A case study of Rotterdam, The Netherlands. *Cities, 25*, 200–212.

Martin-Breen, P., & Anderies, J. M. (2011). *Resilience: A Literature Review (Draft Background Paper)*. Retrieved from http://opendocs.ids.ac.uk/opendocs/handle/123456789/3692.

Molenaar, A., Aerts, J., Dircke, P., & Ikert, M. (2013). *Connecting Delta Cities. Resilient Cities and Climate Adaptation Strategies*. Retrieved from http://www.deltacities.com/.

Pelling, M. (2003). *The Vulnerability of Cities. Natural Disasters and Social Resilience*. London: Earthscan.

Pickett, S. T. A., Cadenasso, M. L., & Grove, J. M. (2004). Resilient cities: Meaning, models, and metaphor for integrating the ecological, socio-economic, and planning realms. *Lanscape and Urban Planning, 69*, 369–384.

Rockefeller Foundation & Arup. (2013). *City Resilience Index. Understanding and measuring city resilience*. Retrieved from https://www.arup.com/publications/research/section/city-resilience-index

Shaw, K. (2012). "Reframing" resilience: Challenges for Planning Theory and Practice *Planning Theory & Practice, 13*(2), 308–312.
Spaans, M., & Waterhout, B. (2016). Building up resilience in cities worldwide – Rotterdam as participant in the 100 Resilient Cities Programme. *Cities.*
Stirling, A. (2014). From Sustainability to Transformation: Dynamics and Diversity in Reflexive Governance of Vulnerability. In A. Hommels, J. Mesman, & W. E. Bijker (Eds.), *Vulnerability in Technological Cultures. New Directions in Research and Governance* (pp. 305–332). Cambridge, MA: MIT Press.
Terenzi, A. (2014). *100 Resilience Cities Setting Agenda – Workshop Report, 26 March 2014.* Archive Gemeente Rotterdam.
The Rockefeller Foundation. (2017). *Cities Taking Action. How the 100RC network is building resilience.* Retrieved from https://medium.com/cities-taking-action
Thrift, N. (2005). Panicsville: Paul Virilio and the Esthetic of Disaster. *Cultural Politics, 1*(3), 337–348.
Vale, L. (2014). The Politics of Resilient Cities: Whose resilience and whose city? *Building Research & Information, 42*(2), 191–201.
Vale, L., & Campanella, T. (Eds.). (2005). *The Resilient City. How Modern Cities Recover from Disaster.* Oxford: Oxford University Press.
Wagenaar, H., & Wilkinson, C. (2013). Enacting Resilience: A Performative Account of Governing for Urban Resilience. *Urban Studies,* 1–20.
Wardekker, A., Jong, A. d., Knoop, J. M., & Sluijs, J. v. d. (2010). Operationalising a resilience approach to adapting an urban delta to uncertain climate changes. *Technological Forecasting & Social Change, 77,* 987–998.
Weichselgartner, J., & Kelman, I. (2015). Geographies of resilience: Challenges and opportunities of a descriptive concept. *Progress in Human Geography, 39*(3), 249–267.
Wilkinson, C. (2012). Social-ecological resilience: insights and issues for planning theory. *Planning Theory, 11*(2), 148–169.

Index

S. Amir (ed.), *The Sociotechnical Constitution of Resilience*,
https://doi.org/10.1007/978-981-10-8509-3

CPSIA information can be obtained
at www.ICGtesting.com
Printed in the USA
LVHW04*2120200518
577859LV00012B/653/P

9 789811 085086